Android Things Quick Start Guide

Build your own smart devices using the Android
Things platform

Raul Portales

BIRMINGHAM - MUMBAI

Android Things Quick Start Guide

Commissioning Editor: Gebin George
Acquisition Editor: Noyonika Das
Content Development Editor: Kirk Dsouza
Technical Editor: Jinesh Topiwala
Copy Editor: Safis Editing
Project Coordinator: Hardik Bhinde
Proofreader: Safis Editing
Indexer: Mariammal Chettiyar
Graphics: Jason Monteiro
Production Coordinator: Deepika Naik

First published: August 2018

Production reference: 1310818

Published by Packt Publishing Ltd.
Livery Place
35 Livery Street
Birmingham
B3 2PB, UK.

ISBN 978-1-78934-179-9

www.packtpub.com

Sometimes, hammers glow for a reason
-Iridessa, about Tinkerbell

To my parents, who always supported me on my endeavours; such as when they gave me my first electronics kit -even if it did not made sense to them as a toy- which planted the seeds for this book.

— Raul Portales

`mapt.io`

Mapt is an online digital library that gives you full access to over 5,000 books and videos, as well as industry leading tools to help you plan your personal development and advance your career. For more information, please visit our website.

Why subscribe?

- Spend less time learning and more time coding with practical eBooks and Videos from over 4,000 industry professionals

- Improve your learning with Skill Plans built especially for you

- Get a free eBook or video every month

- Mapt is fully searchable

- Copy and paste, print, and bookmark content

PacktPub.com

Did you know that Packt offers eBook versions of every book published, with PDF and ePub files available? You can upgrade to the eBook version at `www.PacktPub.com` and as a print book customer, you are entitled to a discount on the eBook copy. Get in touch with us at `service@packtpub.com` for more details.

At `www.PacktPub.com`, you can also read a collection of free technical articles, sign up for a range of free newsletters, and receive exclusive discounts and offers on Packt books and eBooks.

Contributors

About the author

Raul Portales is a software engineer who has had a love for computers, electronics, and gadgets in general for as long as he remembers.

He jumped into Android as soon as it was released. Raul has worked on social networks, education, healthcare, and even founded a gaming studio and a consultancy company. Specializing in mobile and UX, he speaks frequently at conferences and meetups.

Raul's love for electronics reignited when Google announced Android Things. He started tinkering with it with the first Developer Preview, which lead to adding the IoT category on his Google Developer expert profile.

About the reviewer

Gautier Mechling is a software craftsman who's passionate about Android, and a Google Developer expert for IoT.

I would like to personally thank Raul for giving me the opportunity to be among the first readers of this book. It was an interesting and enjoyable read.

Packt is searching for authors like you

If you're interested in becoming an author for Packt, please visit `authors.packtpub.com` and apply today. We have worked with thousands of developers and tech professionals, just like you, to help them share their insight with the global tech community. You can make a general application, apply for a specific hot topic that we are recruiting an author for, or submit your own idea.

Table of Contents

Preface

This book will give you a quick start on Android Things, the platform for IoT made by Google and based on Android. We will go through the basics of IoT and smart devices, interact with a few components that are commonly used on IoT devices, and learn the protocols that work underneath, using examples and a hands-on approach.

We take our hands-on learning approach by going straight into playing with hardware using the Rainbow HAT, so we don't need to do any wiring. We then dig through layer after layer to understand what is being used underneath, but only after we have seen them working. If you are curious about more in-depth learning (such as writing your own drivers), you can always go into the next layer, because almost all the code referenced in this book is open source.

Who this book is for

Since Android Things is a simplified version of Android, you only need a very basic knowledge of how Android works. If you have never done any Android development you will be able to follow along because our examples are kept simple by design.

If you have some experience with other similar platforms, that will be handy, but not necessary. This guide is designed for people that have little to no experience with electronics and microcontrollers but want to get started with them.

Basic knowledge of electronics is desired. That implies that you are familiar with the concepts of voltage and current, as well as resistors, diodes, and capacitors. You also need to know how to read a diagram (there will be a few in this book). All that is very basic knowledge. If you are not familiar with these concepts, you still can follow along, but ultimately you need to understand them to design your own schematics.

Throughout the book we will be using Kotlin as the programming language because it is more modern than Java and allows us to write more concise and easier-to-follow code. Kotlin is meant to be the future of Android development, and it makes a lot of sense to use it on the newest Android platform.

What this book covers

Chapter 1, *Introducing Android Things*, goes over the big picture of Android Things, the vision behind the platform, and how it compares to other – in principle – similar ones, such as Arduino. Then we will explore the Dev Kit, how to install Android Things on them and how connect to the boards to deploy our code to them. Finally, we will create of a project for Android Things using Android Studio, see how it is structured, and what are the differences from a default Android project.

Chapter 2, *The Rainbow HAT*, explains how to use the Rainbow HAT to get started with interactions with hardware. This HAT (Hardware On Top) is a selection of components that you can plug into a Dev Kit in order to start working with them without the need for any wiring. We will learn how to use LED, buttons, read temperature and pressure from a sensor and display it on an LCD alphanumeric display, play around with an RGB LED strip, and even make a simple piano, all using high-level abstraction drivers.

In the next four chapters we will take a look at what is under the hood and we will start exploring the different protocols in depth.

Chapter 3, *GPIO – Digital Input/Output*, goes over **General Purpose Input Output** (**GPIO**), which is what we used for the buttons and LEDs. We will learn how to access them directly and then look at other sensors that use GPIO, such as proximity and smoke detectors, as well as other components that also interact via GPIO, such as relays, DC motor controllers, stepper motors, ultrasound proximity sensors, and a numeric LCD display.

Chapter 4, *PWM – Buzzers, Servos, and Analog Output*, focuses on **Pulse Width Modulation** (**PWM**) and its basic usages, of which we have already have seen the piezo buzzer. The most popular use of PWM is servo motors, so we will see how they work. Finally, we'll learn how to use PWM as an analog output.

Chapter 5, *I2C – Communicating with Other Circuits*, covers the **Inter-Integrated Circuit** (**I2C**) protocol. We have already used it for the temperature/pressure sensor and the LCD alphanumeric display. I2C is the most widely used protocol for simple communication between circuits, and we will explore a few of them. One of the key ones is analog to digital converters (ADC), and we will see how we can use them to read from analog sensors. Other components include magnetometers, accelerometers and IMUs in general, as well as GPIO and PWM expansion boards.

Chapter 6, *SPI – Faster Bidirectional Communication*, is based on the last protocol we'll look into: Serial Parallel Interface (SPI). We have already used this protocol for the RGB LED strip, and in this chapter we will look at some other drivers, such as OLED displays and LED matrix.

Chapter 7, *The Real Power of Android Things*, explores some areas where Android Things really shines by using some libraries and services that enable us to make the most of our developer kit. Among the topics we will cover are the use of Android UI, making companion apps using a REST API, Firebase, and Nearby, and we will briefly explore other libraries, such as Tensorflow for machine learning.

Appendix, *Pinouts diagrams and libraries*, we go over the Pinout diagrams of Rasberry PI and NXP iMX7D. We will then go into details about the state of unsupported Android features and intents on Android Things 1.0, as well as the state of available and unavailable Google APIs on Android Things.

To get the most out of this book

To get the most out of this book you'll need an Android Things Dev Kit and a Rainbow HAT because it will allow you to run all the examples in Chapter 2, *The Rainbow HAT*.

There are a lot of other hardware components that you don't require, but they are interesting to have just to see them working. I recommend that you acquire the ones you are interested in. We go into more detail about developer kits and how to pick the right one, as well as a summary of the other hardware, as part of Chapter 1, *Introducing Android Things*.

Download the example code files

You can download the example code files for this book from your account at www.packtpub.com. If you purchased this book elsewhere, you can visit www.packtpub.com/support and register to have the files emailed directly to you.

You can download the code files by following these steps:

1. Log in or register at www.packtpub.com.
2. Select the **SUPPORT** tab.
3. Click on **Code Downloads & Errata**.
4. Enter the name of the book in the **Search** box and follow the onscreen instructions.

Once the file is downloaded, please make sure that you unzip or extract the folder using the latest version of:

- WinRAR/7-Zip for Windows
- Zipeg/iZip/UnRarX for Mac
- 7-Zip/PeaZip for Linux

The code bundle for the book is also hosted on GitHub at `https://github.com/PacktPublishing/Android-Things-Quick-Start-Guide`. In case there's an update to the code, it will be updated on the existing GitHub repository.

We also have other code bundles from our rich catalog of books and videos available at `https://github.com/PacktPublishing/`. Check them out!

Download the color images

We also provide a PDF file that has color images of the screenshots/diagrams used in this book. You can download it here:
`http://www.packtpub.com/sites/default/files/downloads/AndroidThingsQuickStartGuide_ColorImages.pdf`.

Code in action

Visit the following link to check out videos of the code being run:

`http://bit.ly/2C4VX1M`.

Conventions used

There are a number of text conventions used throughout this book.

`CodeInText`: Indicates code words in text, database table names, folder names, filenames, file extensions, pathnames, dummy URLs, user input, and Twitter handles. Here is an example: "There is also the concept of `AutoCloseable`, which is an interface that all the peripherals implement."

A block of code is set as follows:

```
dependencies {
    [...]
    implementation 'com.google.android.things.contrib:driver-rainbowhat:+'
}
```

When we wish to draw your attention to a particular part of a code block, the relevant lines or items are set in bold:

```
class MainActivity : Activity() {

  private lateinit var led: Gpio

  override fun onCreate(savedInstanceState: Bundle?) {
      super.onCreate(savedInstanceState)
      setup()
      while (true) {
          loop()
      }
  }
}
```

Any command-line input or output is written as follows:

```
$ sudo ./android-things-setup-utility-macos
```

Bold: Indicates a new term, an important word, or words that you see onscreen. For example, words in menus or dialog boxes appear in the text like this. Here is an example: "We select **Install Android Things**."

Warnings or important notes appear like this.

Tips and tricks appear like this.

Get in touch

Feedback from our readers is always welcome.

General feedback: Email feedback@packtpub.com and mention the book title in the subject of your message. If you have questions about any aspect of this book, please email us at questions@packtpub.com.

Errata: Although we have taken every care to ensure the accuracy of our content, mistakes do happen. If you have found a mistake in this book, we would be grateful if you would report this to us. Please visit www.packtpub.com/submit-errata, selecting your book, clicking on the Errata Submission Form link, and entering the details.

Piracy: If you come across any illegal copies of our works in any form on the Internet, we would be grateful if you would provide us with the location address or website name. Please contact us at copyright@packtpub.com with a link to the material.

If you are interested in becoming an author: If there is a topic that you have expertise in and you are interested in either writing or contributing to a book, please visit authors.packtpub.com.

Reviews

Please leave a review. Once you have read and used this book, why not leave a review on the site that you purchased it from? Potential readers can then see and use your unbiased opinion to make purchase decisions, we at Packt can understand what you think about our products, and our authors can see your feedback on their book. Thank you!

For more information about Packt, please visit packtpub.com.

Introducing Android Things

Welcome to *Android Things Quick Start Guide*. In this introductory chapter, we will look at the big picture of Android Thing, how it compares to other—in principle—similar platforms, such as Arduino, and explore the vision behind the platform. We will check the different developer kits available exploring the pros and cons of each one. We will also look at the other components we will use throughout the book. We will also learn how to install Android Things on a developer kit and how to connect to it from our development computer in order to deploy our code. Finally, we will follow the creation of a basic project for Android Things using Android Studio, see how it is structured, and discuss the differences with a default Android project. The topics covered are as follows:

- What is Android Things?
- Design concepts behind Android Things
- Hardware you will need
- Setting up a developer kit
- Creating an Android Things project

So, let's get started.

Technical requirements

You will be required to have Android Studio and Android Things installed on a developer kit. You also will require many hardware components to effectively perform the tasks given in this chapter. The components are very interesting to have, just to see them working, but the Rainbow HAT is particularly important. We go into details about the developer kits and how to pick the right one, as a part of Chapter 1, *Introducing Android Things*. Finally, to use the Git repository of this book, you need to install Git.

The code files of this chapter can be found on GitHub:
https://github.com/PacktPublishing/Android-Things-Quick-Start-Guide.

Check out the following video to see the code in action:

```
http://bit.ly/2wC0tyR.
```

What is Android Things?

Android Things is the IoT platform made by Google and based on Android. You could have guessed that by its name. It is similar to Android Wear, Android Auto, and Android TV in the way that it is an adaptation of Android to another domain, where most of the concepts of the platform are still valid, but there are significant differences as well.

The platform is intended to design and build IoT devices and bring them to the mass market. The key idea behind it is that you can easily prototype your project and, once it is ready, you can move from your developer kit to a simpler and smaller carrier board with just the **SoM** (**System-on-Module**).

SoMs are very handy. They are a step up from **SoCs** (**System-on-Chips**). They integrate RAM, flash storage, Wi-Fi, and Bluetooth on a single module. The official boards come with the FCC certifications, so all the process of getting the software onto them is streamlined.

Throughout the process of going to mass market, Google will provide security updates to the platform, so keeping your IoT devices up to date and secure is no longer something you have to worry about.

 Google will provide security updates to Android Things.

All the areas of this process are centralized on the Android Things Console, which has a similar function to the Google Play Console, but is more focused on building images that you can then distribute and deploy.

One of the key advantages of Android Things is that it can make use of almost all of the already existing Android libraries and frameworks, which gives it a head start in terms of tooling. From the developer point of view, it also lowers the entry barrier to make IoT devices, since all the knowledge you have of the Android framework, libraries, and tools is fully transferable.

In simple terms, Android Things simplifies and empowers the development of IoT devices.

Internet of Things (IoT) vs. smart devices

Before we go any further, let's try to establish what most people understand by IoT and what smart objects are.

An IoT device is meant to use the internet as something integral for its behavior. Examples include remote controlled lights, any appliance that you can interact with over an API (such as lights, ovens, thermostats, coffee makers, and so on) even if it is only for reading data, such as an umbrella that glows when there is a rain forecast, or writing it, such as a simple weather station. For all those systems, the internet is an integral part of the concept.

On the other hand, a smart object is something that is designed to work in isolation. This can still include many complex components, such as machine learning and computer vision. A self-driving robot that navigates a maze or smart curtains that open when there is sunlight are both quite autonomous. A device that you show a photo of a dog to and tells you the breed also falls into this category (and yes, that is one of the official demos of Android Things). All these examples are smart, but they do not need the internet, so I like to call them just smart devices and, yes, we can build those with Android Things too.

Android Things vs. other platforms

Probably the most commonly asked question is: how does Android Things position itself with Arduino?

They are completely different platforms. Android Things is a full operating system running on an SoM, while Arduino runs on microcontrollers. From a capabilities point of view, they are in different orders of magnitude.

Everything you can do on Arduino, you can do with Android Things, but the opposite is not true; there are many things you can do on Android Things that you can't even get close to with Arduino, and we will cover some of them in the final chapter.

Lastly, there is the comparison of SDK and tools. Android Studio is one of the best IDEs overall. The Arduino IDE is very simple and limited. Not to mention how much easier it is to manage project dependencies on Android Studio compared to Arduino libraries.

However, not everything is better on Android Things. There are a few areas where Arduino has the upper hand, namely, power consumption and analog input/output.

The power consumption of Arduino is very low; you can run it on batteries or even solar power for extended periods of time. That is definitely not possible with the current developer kits for Android Things; even a large battery pack will get depleted in just a matter of days.

There is room for everything on IoT. I think Arduino is best at having sensor data collection on the field, but then a central hub running Android Things can do aggregation, cloud upload, and even use machine learning to interpret the readings.

The other platform I usually get asked to compare Android Things with is a Raspberry Pi running Linux and programming it on Python. In this case the hardware is exactly the same, but there are two main advantages of Android Things.

Firstly, if you plan to eventually release and sell your IoT project, Android Things offers a developer console and even a means to mass produce devices, as well as automatic security updates.

Secondly, even with all the libraries that are available on Python, there are more and better examples of using Android. Building an IoT app is a lot closer to a mobile app than it is to a desktop app, so all your prior Android knowledge is easily transferable to Android Things.

Android Things shines when using all the tools that it has available; be it services such as Firebase for cloud storage or cloud messaging and TensorFlow for image classifiers, or libraries such as Retrofit to create API clients and NanoHTTPD to create a server, or just SDK classes, such as `Thread`, `Alarm`, `Timer`, and so on.

Emulators and testing

Since Android Things is meant to be used with custom hardware, it is useless to have an emulator of just the dev kit. Also, given the wide variety of hardware that we can connect, it is almost impossible to emulate it; we would need a whole electronics set to be able to draw our project and then run it. Those are the reasons why Android Things does not have an emulator.

 There is no emulator for Android Things.

In my experience, the lack of an emulator is not a problem. On the one hand, it is very satisfying to see the hardware working, and, on the other, many problems that you'll face will not be solved with an emulator, problems such as deciding how to arrange the components inside the chassis of a robot car or figuring out how many turns of a stepper motor are needed to activate a candy dispenser.

In my experience, the most common source of bugs when working with electronics is the actual wiring of the circuits. A loose cable or a misplaced connection are often the reason why something is not working and an emulator won't solve that. Besides, setting the connections on a program is usually more time-consuming than actually wiring them.

Finally, if you really want to use something like an emulator to test your project as you go along, my suggestion is to rely on testing using a mocking framework, such as Mockito or EasyMock, and either run instrumentation tests on the device or use Robolectric to run them on your computer.

 You can test Android Things projects using Mockito and Robolectric.

While testing on Android Things is beyond the scope of this guide, and I don't recommend investing in it for just prototyping and tinkering; if you are planning on going to the mass market with any IoT project, testing should definitely be on your agenda.

Android Studio

One of the big advantages of Android Things is that it relies on the excellent set of tools that Google has been building for Android over the years. Among those tools, Android Studio is probably the most impressive.

It is not only that you do not need to get used to a new IDE. Android Studio is already an excellent tool.

You have dependency management with gradle, multi-module and library projects, support for flavors, and so on. And that is just basic gradle. The IDE itself has excellent wizards, refactoring tools, search and autocompilation, profiling, and much more. Plus, do not forget Kotlin support.

Android Studio has come a long way since the early days of Android with Eclipse, and Android Things has all those tools at its disposal from day one.

Design concepts behind Android Things

We have talked about the vision that drives Android Things. Let's now look at the design considerations as a platform that are reflected in how it works and the options it gives to us, namely being designed without the requirement to have a display, the meaning of Activity and Home, and a different way of working with permissions and notifications.

Displays are optional

The most significant characteristic of Android Things is that it has been designed to work without a graphical user interface. This is a very logical decision if you consider that most IoT devices do not have a screen. It is important because it dictates that all the APIs that have a strong requirement for a screen need to be redesigned or disabled/removed.

Quite often, this design decision is misunderstood as that Android Things does not support graphical user interfaces. That is not correct. You can use the same classes as you would do when building a UI on Android, the same way to create layouts and assign them to an activity. Those layouts will work fine and you can rely on them if you are building a device that has that sort of interface, but it is not a requirement for an Android Things project.

 If you want to build a UI for Android Things, you can do it in the same way you would do it on Android.

One small difference is that Android Things does not have a status bar or navigation buttons and, therefore, the application window takes all the real state of the display. This also implies that there is no way of displaying notifications on Android Things, so the Notification API is not present

 Although you cannot display notifications on Android Things, you can use Firebase Cloud Messaging to send messages to your app to trigger events.

Finally, even though you may not have a UI, the key component we will use to build our applications is the `Activity`, but, in many cases, it will not have a visual component.

Home activity support

Similar to Android home screen launchers, Android Things supports the concept of home, and it is defined the same way. This is the application that will run when the device boots. We will see how this is declared later in the chapter when we explore the default project of Android Things.

An IoT device will usually have a single purpose, in contrast with mobile phones that have multiple apps that can be launched. That single purpose is what the Home application does.

 IoT devices usually have a single purpose. We use the Home application to define it.

While developing, you can launch apps in the same way you can on Android phones, also having them declared on the `Manifest`. This is useful for quickly testing some ideas, especially if you are just playing around with simple apps and do not want them to become the home app, but keep in mind that you won't have the ability to launch them in any other way.

Permissions

Granting permissions to apps is a good example of a feature that has been redesigned because having a display is not mandatory. In Android, permission requests are pop ups that ask the user. So how do we do this without a user interface?

Firstly, the permissions are declared in the same way as you would do for an Android app; you just add them to your `Manifest`.

Then, permissions are granted at install time, either with Android Studio when you are developing, or via the Android Things Console if you include the app for distribution.

You can also use `adb` to grant and revoke permissions

Android Things Console

Android Things provides a console that has a similar role to the Google Play Console for mobile apps, although its functionalities are quite different.

The key functionalities that the console provides to developers are:

- Downloading and installing the latest Android Things system image
- Building factory images that contain OEM applications along with the system image
- Pushing **over-the-air** (**OTA**) updates
- Managing OEM applications
- Analytics

The Android Things Console will help us with distribution.

It is intended to be used when you move to mass production and you want to build the image to be flashed on your boards.

Given the nature of this book, the console is not something we will be exploring. We will just use it to download the script that flashes the default image on the developer kits.

Supported APIs

Android Things relies heavily on Android libraries to offer flexibility and functionality. Given that Android Things is optimized for embedded devices that may not contain the same feature set as an Android phone or tablet, not all the APIs are present. In most cases this is coming from the restriction of not requiring a UI. There is a reference list of system APIs and Google APIs that are supported and unsupported in the Appendix.

You can use `hasSystemFeature()` to determine whether a given device feature is supported at runtime.

As a rule of thumb, if it requires a system service -such as a calendar or telephone- or a strong UI component -such as AdMob or Sign-in- it is likely to not be supported, but the number of available services is remarkably large.

Hardware you will need

As already mentioned, the Android Things SDK does not include an emulator . If you want to test your code, you need to get a developer kit. There are only two supported developer kits, which are Raspberry Pi 3 and NXP Pico iMX7D. There are more boards available for production (such as Qualcomm SDA212 and MediaTek MT8516), and there were other developer boards during the developer preview phase (Intel Curie and Edison early on, and two other NXP boards more recently), but they are all deprecated.

To check the latest information about developer kits and supported platforms, visit `https://developer.android.com/things/hardware/`.

We will look at the differences of both boards so that you can pick the one that suits you best.

You will also need a Rainbow HAT for all the examples on the next chapter and, in part, for the rest of the book. It will allow you to get hands-on quickly by letting you focus on writing software that interacts with hardware without the need for any wiring. The Rainbow HAT was designed for Android Things in particular and all developer kits have an option to include it.

As we advance through the book and we start interacting with other hardware; you will need some breakout circuits to be able to test the code. I have picked components that are relatively cheap and are supported on Android Things. We'll have a quick overview in this section. You do not need to get them all, but they are fun to play with.

Most likely, you will not need any breadboards, resistors, and suchlike, although they are sometimes necessary. In any case, you will still need some wires. Dupont connectors are very handy, especially female-female ones in order to connect the dev kit to the other components.

Android Things developer kits

As we mentioned, there are two developer kits available; one with a Raspberry Pi and one with an NXP iMX7D.

You can always just get a supported board or, if you already have a Raspberry Pi, repurpose it; but the kits come with some added niceties.

Raspberry Pi

The key advantage of a Raspberry Pi is that most people already have one, so getting started on Android Things is very easy.

The developer kit includes a special case that makes it visually appealing, but that's not just it, it includes the name and function of each pin, which turns out to be very handy.

Other advantages of this developer kit are as follows:

- HDMI connector
- Four USB ports
- Price

However, this board has one big disadvantage: the micro-USB port is for power only, so it can only be accessed with `adb` via a network connection.

The Raspberry Pi micro-USB port is for power only, cannot be used for data.

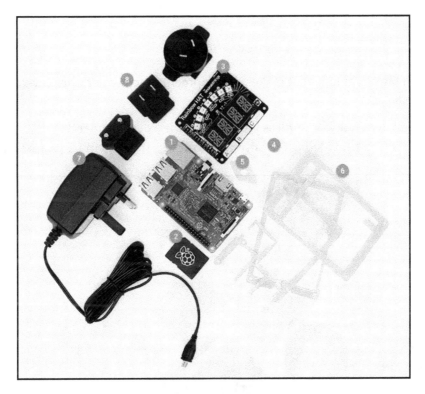

Raspberry Pi developer kit overview

If you prefer having Wi-Fi and HDMI out of the box (or if you already have one), pick the Raspberry Pi.

NXP Pico iMX7D

The second developer kit is way more complete. It includes the following:

- 5" touch screen (and connector)
- Camera (and connector)

The main advantage of this kit, however, is not the extra hardware; it is that you can use `adb` (the command-line tool) over the USB type C cable in the same way you would do for Android on a phone.

 You can use `adb` over the USB on the iMX7D.

While that is a big advantage that makes development much easier, this developer kit has some drawbacks:

- Wi-Fi does not work unless you connect the external antenna
- The screen provided is the only way to access the system UI (there is no HDMI)

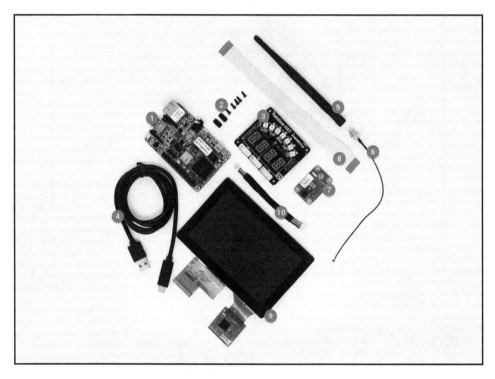

NXP Pico iMX7D developer kit overview

If you prefer USB deployment and debugging and don't mind the price tag, NXP should be your choice.

Rainbow HAT

The **Rainbow HAT** (**Hardware On Top**) is a component you place on top of your developer kit—as with any other HAT—and is specifically designed to give a quick start into Android Things. It contains (in the maker's own words) *a buffet of sensors, inputs, and displays to explore Android Things.*

Using the Rainbow HAT has many benefits. It allows us to focus on the software side of Android Things, not having to worry about connecting anything, wiring, pinouts, and protocols. The selection of hardware is very good and it covers a wide range of components that work in different ways.

You can purchase it as a standalone component, but most Android Things developer kit bundles include it.

We will be using the Rainbow HAT to interact with hardware in an easy way

Although it was originally designed for the Raspberry Pi, the pinout is compatible with the iMX7D, so it can be used on both.

 The Rainbow HAT works on both Raspberry Pi and iMX7D.

The Rainbow HAT includes:

- Three single color LEDs, in red, green, and blue, above each button
- Three capacitive buttons (labeled A, B, and C)
- A piezo buzzer
- A four-digit 15-segment LCD alphanumeric display
- BMP280 temperature and pressure sensor
- Seven APA102 multi colour LEDs
- Some extra pins, labeled

In the following chapter, we will explore each and every one of the components with simple programs to get familiar with the usage of Android Things drivers.

In later chapters, we will still use the Rainbow HAT at the beginning as a means to get familiar with the underlying protocols that are used for each component, in order to then move onto other components.

Components, circuits, and so on

In most examples, we will use components that have drivers supported by Google as `contrib-drivers`. The remaining ones will be collected from other developers, including my own repository, PlattyThings.

The following is a list of the different components that we will be using across our examples. For the ones that have a particular controller chip, I have added the code of the controller so that you can make sure you have a compatible component:

- 3v3 relay
- Pyroelectric Infrared PIR Motion Sensor
- MQ135: smoke sensor
- L298N: dual DC Motor controller (and two DC motors)
- ULN2003: stepper motor controller (28YBJ-48 – Stepper motor)
- HC-SR04: ultrasound proximity sensor
- M1637: a seven-segment, four digit LCD
- Tower Pro MG90S: servo motors
- LEDs (single color and RGB)
- PCF8591: Analog-to-Digital Converter (ADC)
- PCA9685: PWM extension board
- PCF8575: GPIO extension board
- LCD display with LCM1602 controller
- MPU6050: Gyroscope
- MAX7219: LED matrix
- SSD1306: OLED display (wired to I2C and SPI)

Each of the chips usually requires some basic setup wiring that involves a few resistors. To avoid having to wire all that into a breadboard, we will be using breakout circuits, which are basically a small printed board with the chip and everything that is required, so it can be connected directly to our board.

 Using breakout circuits is a good way to simplify wiring.

However, in a few cases we will still need to use breadboards. Essentially, this is the case when using buttons and LEDs, which are not worth putting on a breakout circuit just themselves.

Wires and breadboards

The last components we will need are cables and breadboards. The most convenient cables have Dupont connectors. Dupont female connectors can be inserted on the pins on the boards, or in the breakout circuits, while male connectors can be inserted on breadboards.

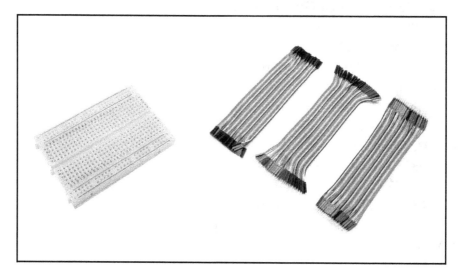

Breadboard and Dupont cables

We will be using female-to-female Dupont connectors most of the time, but some male-to-female ones will be useful to connect to a common power source or ground.

 Dupont connectors are the best way to wire prototypes. Female-to-female connectors are especially useful.

When we are making prototype circuits, we usually do them on breadboards. Each column of a breadboard is connected, and so are the top and bottom rows (which are generally used for Vcc and ground)

If you have some LEDs, buttons, and resistors, you can make use of them, but, in any case, we will keep wiring on breadboards to a minimum. Some interesting components that you can use are variable resistors, such as potentiometers, light resistors, and thermistors.

There are several **starter packs** typically designed for Arduino that combine most of these components and they are always handy to have around.

Finally, you can get your hands on a multimeter. We won't need one for the examples in this book, but it will be very useful if you plan to get more hands-on with electronics.

Setting up a developer kit

To configure our Android Things developer kit, we will use the setup utility script that can be downloaded from the Android Things Console. It is a simple wrapper of a few tools and simplifies the process of getting our board ready.

The setup utility will help us to install Android Things on the developer kit and also to configure Wi-Fi.

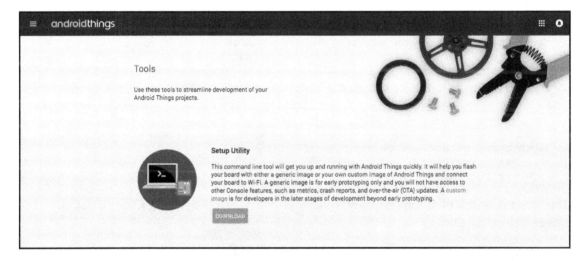

You can download the setup utility from the Android Things Console

The download contains executables for Linux, macOS, and Windows. If you are going to flash a Raspberry Pi, make sure to run it with administrator permissions (sudo on macOS/Linux) because writing the SD card requires root permissions.

Installing Android Things

To proceed with the installation of Android Things, we just have to execute the script for our operating system:

```
$ sudo ./android-things-setup-utility-macos
```

If you are flashing a Raspberry Pi, you will need to run the script with sudo.

The script is quite self-explanatory. We just have to follow the steps and select what we want to do. The trickiest part is to configure Wi-Fi, but we can do that later any time in several ways. It looks like this:

```
Android Things Setup Utility (version 1.0.19)
==============================
This tool will help you install Android Things on your board and set up Wi-
Fi.
What do you want to do?

1 - Install Android Things and optionally set up Wi-Fi
2 - Set up Wi-Fi on an existing Android Things device
```

We select **Install Android Things**. We will be asked about the hardware since there are differences in the process. For this example we will be using a Raspberry Pi because it is a bit more complex. The setup for the iMX7D is simpler, since you do not need to extract the SD card or connect to the network via a cable.

```
What hardware are you using?
1 - Raspberry Pi 3
2 - NXP Pico i.MX7D
```

The script will download the required tools and will ask us if we want to use a default image or a custom one.

 To get started on Android Things, a default image is the best option.

For our purposes, and for development in general, a default image is the best option. The custom image is used when you are later in the process of building your device and want to test that everything works as expected:

```
Raspberry Pi 3
Do you want to use the default image or a custom image?
1 - Default image: Used for development purposes. No access to the Android
Things Console features such as metrics, crash reports, and OTA updates.
2 - Custom image: Upload your custom image for full device development and
management with all Android Things Console features.
```

The script will then download the required image and flash it into the device.

If you are using an NXP board, this is done via the USB type C cable. Please note that the script is a bit confusing in terms of whether you have to connect to your laptop directly or via a USB hub. In any case, make sure the port can properly power the device. This is likely to fail if you use a non-powered USB hub.

 When setting up an iMX7D, make sure the USB port can power the device properly.

```
Connect your device to this computer:
The USB cable should plug into your board's USB-C port. If your computer
also has USB-C ports like the more recent MacBooks, you will need to use a
USB hub.
Otherwise the board won't power on correctly.
```

In the case of a Raspberry Pi, you need to extract the SD card and insert it in your computer so that it can be flashed. This is the step that requires root permissions.

```
Downloading Etcher-cli, a tool to flash your SD card...
20.5 MB/20.5 MB
Unzipping Etcher-cli...

Plug the SD card into your computer. Press [Enter] when ready
[...]
iot_rpi3.img was successfully written to Apple SDXC Reader Media
(/dev/disk3)
Checksum: 4d176685
```

And with that, our board has Android Things installed. Now, we have the option to configure Wi-Fi, which we are going to do in the next section.

```
If you have successfully installed Android Things on your SD card, you can
now put the SD card into the Raspberry Pi and power it up. Otherwise you
can abort and run the tool again.
```

This last message is a bit confusing. It asks us whether we have successfully installed Android Things. It refers to the need for root permissions to write on the SD card. If you forgot to run the script with sudo, it will fail and you'll need to start over again.

Configuring Wi-Fi using the script

We already mentioned that you can only deploy your code to a Raspberry Pi using `adb` over a network connection, so configuring Wi-Fi is quite important. I do use an Ethernet cable most of the time when testing on a Pi, but configuring Wi-Fi comes in handy quite often.

In the case of the iMX7D, this step is even less important, since we can use the board for debugging via the USB type C cable.

In any case, there are several ways to configure Wi-Fi: you can use the script any time to perform just this step, you can do it with the `adb` command (for which the script is a wrapper) or you can use the system UI for that.

Let's take a quick look at the message from the script:

```
Would you like to set up Wi-Fi on this device? (y/n)
Please plug your Raspberry Pi to your router with an Ethernet cable, then
press [Enter].
```

So, to be able to configure Wi-Fi on a Raspberry Pi, you need to connect it via Ethernet first; then, the script will connect to it using `adb` and configure Wi-Fi.

 You need an Ethernet cable to configure Wi-Fi on the Raspberry Pi with the setup script.

In the case of the iMX7D, the message is slightly different:

```
Please ensure antenna is already attached. If it is not, disconnect your
board, attach the antenna and reconnect your board to your computer.

When ready, press [Enter]...
```

This board will not be able to connect to Wi-Fi unless the antenna is connected, but the configuration is done via the USB type C cable:

 The Wi-Fi on iMX7D will not work unless the antenna is connected.

```
Attempting to connect to your Raspberry Pi at Android.local...
Connected to device through Ethernet.
Enter the Wi-Fi network name: PlattySoftHQ
Enter the Wi-Fi network password (leave empty if no password):
Connecting to Wi-Fi network PlattySoftHQ...
Looking for device... This can take up to 3 minutes.
Device found.
Waiting...
```

What the script does is to connect `adb` to the device using the name `Android.local` and then send an `adb` command to set up the Wi-Fi. If you want to do that by hand, you just have to type the following:

```
$ adb connect Android.local
connected to Android.local:5555
$ adb shell am startservice \
    -n com.google.wifisetup/.WifiSetupService \
    -a WifiSetupService.Connect \
    -e ssid <Network_SSID> \
    -e passphrase <Network_Passcode>
```

You can see the complete list of parameters in the official documentation at `https://developer.android.com/things/hardware/wifi-adb`.

 The Wi-Fi setup may fail if you have several Android devices on the LAN.

Since the script relies on the device identifying itself as `Android.local` on the local network, this only works if it is the only device doing that. If you already have another device on the network that uses that name (such as another Android Things or an Android TV) this may fail, so let's look at a back-up plan.

Configuring Wi-Fi using the system UI

Early versions of Android Things did not have any UI besides a screen that displayed the Android Things logo and the current IP. Part of the feedback received was that using `Android.local` as the means to connect to the board was way too unreliable and, since you need to connect a display to the dev kit to see the IP if that fails, you could also configure it straight away, so that feedback was incorporated to the default app.

On the current version of Android Things, you can configure Wi-Fi the same way you would on an Android device. On a Raspberry Pi, you want to connect an HDMI display, mouse, and keyboard (or a touchscreen, if you have one). The iMX7D should work without any issues with the script, but you can also use the touchscreen that is provided with the dev kit.

The main screen of Android Things is similar to an Android settings screen, but with a far fewer options. Let's pick networks and then enable wireless.

 I recommend configuring DHCP to always give the same IP to these boards and to write it down, so you don't have to connect a display every time you start the device to check the IP.

Home screen of Android Things, including information and settings

By clicking on a **Wi-Fi**, you can just enter a password directly on the device.

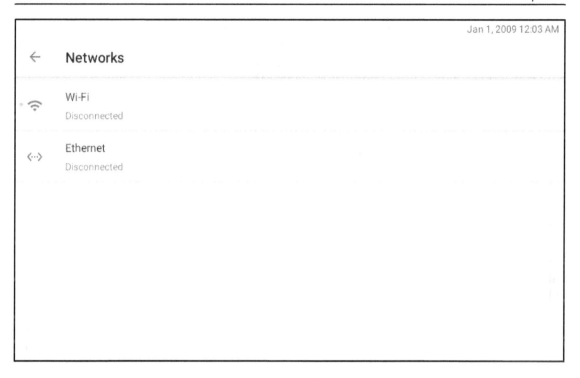

Network configuration screen of Android Things

You should take this chance, now that you have a screen connected, and write down the IP for later use.

Connecting to your Android Things

On an iMX7D, you just need to use `adb` as you would do with an Android-powered smartphone; no extra setup is required.

 You do not need to perform any extra steps to connect to an iMX7D.

For Raspberry Pi, you need to connect to `adb` over the network, so, regardless of whether it is connected via a cable or via Wi-Fi, you can just type the following:

```
$ adb connect Android.local
```

If you are lucky that should be enough, but if, as we mentioned before, you have several devices identifying as `Android.local`, this will fail. In that case, you will need to connect using the IP of the dev kit. The simplest way to know the IP address of a device is to attach a display to it, so hopefully you already have that IP written down.

 The simplest way to know the IP address of a dev kit is to attach a display to it.

Once you know the IP (let's use the one for the screenshot: **192.168.42.52**), you can connect to the board by typing the following:

```
$ adb connect 192.168.42.38
connected to 192.168.42.38:5555
```

And you are ready to deploy your apps to the dev kit. It will appear with a name, such as **Google Iot_rpi3,** on the **Select Deployment Target** dialog of Android Studio.

Now that we have our development kit ready to be used, let's create a sample empty project for Android Things.

Creating an Android Things project

Creating an Android Things project is very easy. Android Studio has included it as a form factor inside the default wizard for **New Project**.

So, let's go ahead and create a project. When we are asked about the target devices, we unselect **Phone and Tablet** and select **Android Things** instead:

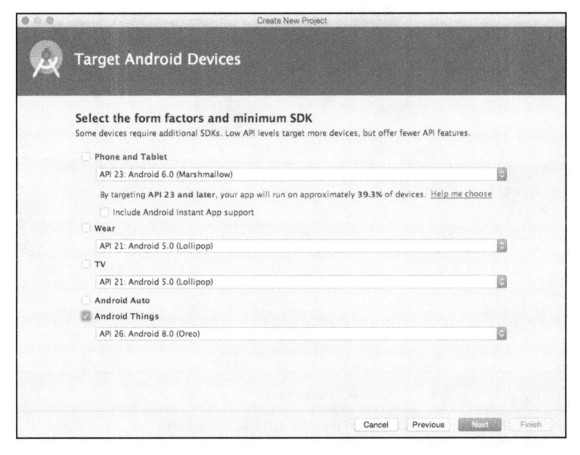

We can select Android Things as a form factor

The next step is to choose whether we want to have an Activity added to our project and, if so, of which kind. Here we have the first important difference because we have the option of an **Empty Activity** as well as a **Peripheral Activity**:

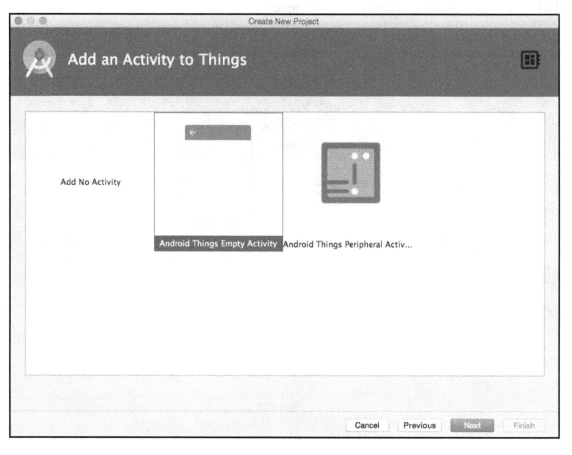

Several options to Add an Activity to Things

The main difference between the two types of activities is that in the case of **Peripheral Activity**, the last step allows us to configure several peripherals that are included in the `contrib-drivers` package and will insert initialization code for such devices in the Activity as well as the dependencies in `gradle`.

Another small difference is that an **Empty Activity** will add an `onCreate` method, while a **Peripheral Activity** (even if we do not add any drivers) will add both `onCreate` and `onDestroy`.

For now, we will create a **Peripheral Activity** with no peripherals.

Don't forget to select **Kotlin** as programming language. We will be using Kotlin because it is more modern than Java and allows us to write code that is more expressive and easier to follow. Kotlin is also meant to be the future of Android development and it makes a lot of sense to use it in the newest Android-based platform.

Android Studio also offers the possibility of creating an Android Things module to be added to an existing project. This is interesting when you are building an IoT device with a companion app. You can have a mobile module and an Android Things module, both inside the same project.

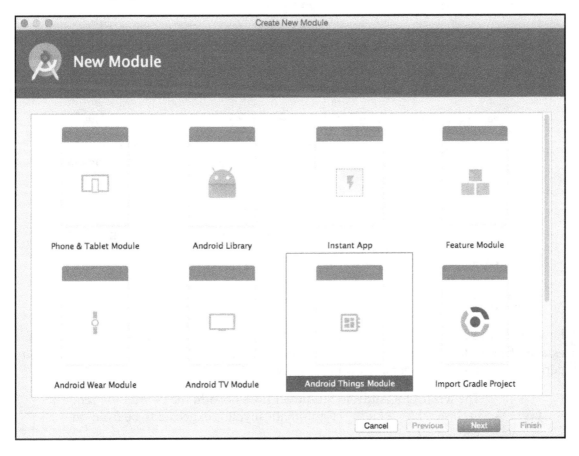

We can create an Android Things module at any time

Android Things projects have a few particularities that distinguish them from the standard Android ones. Let's take a look at the code that has been generated and explore the differences.

Manifest

The place with most differences is the `Manifest`.

```xml
<?xml version="1.0" encoding="utf-8"?>
<manifest xmlns:android="http://schemas.android.com/apk/res/android"
 package="com.plattysoft.myapplication">

    <uses-permission
android:name="com.google.android.things.permission.USE_PERIPHERAL_IO" />

    <application>
        <uses-library android:name="com.google.android.things" />

        <activity android:name=".MainActivity">
            <intent-filter>
                <action android:name="android.intent.action.MAIN" />
                <category android:name="android.intent.category.LAUNCHER" />
            </intent-filter>
            <intent-filter>
                <action android:name="android.intent.action.MAIN" />
                <category android:name="android.intent.category.HOME"/>
                <category android:name="android.intent.category.DEFAULT"/>
            </intent-filter>
        </activity>

    </application>
</manifest>
```

First, we have a new permission request: `USE_PERIPHERAL_IO`. This permission does not exist in the standard Android SDK; it is specific to Android Things and is required to open peripherals, which are essentially everything we will be working with.

> The wizard for Android Studio 3.1 does not add the permission request for `USE_PERIPHERAL_IO` automatically. It is expected to be fixed on version 3.2.

This permission was added on the release candidate, so many hobbyist projects that were created for earlier versions of Android Things do not have it. Pay attention if you are importing samples or drivers from the community, as this permission may be missing.

 Early developer previews of Android Things did not require this permission, so be careful if you are looking at the source of old projects

Inside the application tag, we see a `uses-library` declaration. This indicates a dependency on a library to allow installation. If the device does not have the Android Things library, it will not let us install the app on it. This is similar to what happens with the `leanback` library of Android TV. The library is part of the Android Things operating system, so we do not have to do anything; that code is there to prevent installation of IoT apps on phones by mistake.

A third difference is that the activity we have created has two intent filters, one for `LAUNCHER` and one for `HOME`. A normal Android app will not have the `HOME` category. As we mentioned earlier, Android Things is designed in a way that it launches a single application, and that application is marked with the category `HOME`. Earlier versions used the category `IOT_LAUNCHER`, but it was replaced with `HOME` on the release candidate.

The reason behind having a second intent filter is simply to allow us to launch the application from Android Studio. By marking an Activity with the category `LAUNCHER`, we can run it from the IDE. This simplifies development significantly, especially if you reuse the same board for several projects.

 Early developer previews of Android Things used the category `IOT_LAUNCHER` instead of `HOME`. You may still find it on outdated projects.

Gradle configuration

The second area that has some differences is the gradle configuration. Let's look at `build.gradle`.

Besides the dependencies and configuration to allow the use of Kotlin, we can see that a new library has been added to the dependencies:

```
compileOnly 'com.google.android.things:androidthings:+'
```

This is the counterpart of the library we have declared on the `Manifest`. The gradle dependency is marked as `compileOnly`, which means that the library is expected to be installed on the target device and will not be included in the code of the application.

Activity code

We can see that the `Activity` has just two methods, `onCreate` and `onDestroy`.

```
class MainActivity : Activity() {

    override fun onCreate(savedInstanceState: Bundle?) {
        super.onCreate(savedInstanceState)
    }

    override fun onDestroy() {
        super.onDestroy()
    }
}
```

It is noticeable that both methods are empty; while we are used to setting the layout using `setContentView` on standard Android applications, that code is not there. The explanation is, one again, that displays are optional on Android Things and, therefore, the creation of a layout is optional.

The second interesting point is that there are two methods. This is meant to enforce the lifecycle of peripherals.

The recommended way to work with peripherals is to open them inside `onCreate`, keep a reference to use them along the app, and close and release them inside `onDestroy`. Remember, peripherals are hardware that is still on once your application finishes, so if one app leaves them in a particular state, you'll find them in the same state. If an app does not release them, you may not be able to open them and get an `IOException`.

 We will place the initialization of peripherals inside `onCreate` and we will release them inside `onDestroy`.

You can, however, open a peripheral, perform some operations with it, and close it straight away. It is less efficient but, in certain cases, can be handy.

Most peripheral drivers implement `AutoCloseable`, so they will close themselves even if you forget to do so once the application is destroyed, but it is best to stick to the recommended practices.

 Many community drivers have not been updated to fix the breaking changes on Android Things Developer Preview 7 and 8.

Throughout this book, we will be using drivers from the Google-maintained `contrib-drivers` repository and also from other community-maintained ones, although most will be from PlattyThings. On Android Things Developer Preview 7 and 8, breaking changes were introduced to the SDK, and many community drivers have not been updated to the latest version. I have made that repository to ensure that all the drivers work on Android Things 1.0.

Summary

In this first chapter we have learned about the big picture of Android Things, the vision behind the platform, and how it compares to other IoT platforms. We explored the different developer kits available so you can pick the one that suits you best and learned how to install Android Things on them and how to connect them to our development computer. Finally, we followed the creation of a basic project for Android Things with Android Studio, saw how it is structured, and looked at the differences with a default Android project.

Now that we have everything ready, it is time to start tinkering with the Rainbow HAT, which will be the focus of our next chapter.

2
The Rainbow HAT

In this chapter, we will make use of the Rainbow HAT to learn how to interact with hardware. This **HAT** (**Hardware on Top**) is a selection of components that you can plug into a developer kit to start working with them without the need for any wiring. We will start with an overview of the architecture of Android Things—explaining peripheral lifecycle and user space drivers—and then work with each component of the Rainbow HAT (as shown in the following screenshot) including some best practices along the way.

The Rainbow HAT with all the LEDs and LCD on

For each part we will use a complete self-contained example and we will analyze how the code works. Along this chapter, we will be using the high-level abstraction drivers for the components that come from the `contrib-drivers` repository from Google.

We will also take advantage of the meta driver for the Rainbow HAT that makes accessing this hardware even easier. It includes the drivers for all the components of the Rainbow HAT plus some utility methods to access the peripherals. So, to get started, you only need to add one line to the `gradle` dependencies of your project:

 The Rainbow HAT driver includes all the other required drivers, so we don't need to include anything else.

```
dependencies {
    [...]
    implementation 'com.google.android.things.contrib:driver-rainbowhat:+'
}
```

At the end of the chapter we will take a quick look at what is under the hood and we will learn which protocols we have been using underneath, which then will be explored in following chapters, using the concepts of this one as a foundation.

The topics covered in this chapter are as follows:

- Android Things architecture
- LEDs
- Buttons
- Piezo buzzer
- Alphanumeric display (Ht16k33)
- Temperature and pressure sensor (Bmx280)
- LED strip (Apa 102)

Technical requirements

You will be required to have Android Studio and Android Things installed on a developer kit. You also will require many hardware components to effectively perform the tasks given in this chapter. The components are very interesting to have, just to see them working, but the Rainbow HAT is particularly important. We go into details about the developer kits and how to pick the right one, as a part of `Chapter 1`, *Introducing Android Things*. Finally, to use the Git repository of this book, you need to install Git.

The code files of this chapter can be found on GitHub:
https://github.com/PacktPublishing/Android-Things-Quick-Start-Guide.

Check out the following video to see the code in action:

```
http://bit.ly/2NajYsI.
```

Android Things architecture

Before we get our hands on into the code, let's explore the overall architecture of Android Things and present a few concepts that we will be using throughout the chapter, such as the best practices to handle the peripheral lifecycle and the concept of user space drivers.

Peripheral lifecycle

Whenever we want to use a peripheral, we need to get access to it before we can do anything with it. This is done using the `open` method that all peripheral classes have. Opening a peripheral gives us exclusive control over it and therefore it can not be opened if it is already open by another app. The operating system will throw an `Exception` if the peripheral is already in use.

As you can imagine, this also means that we need to release the peripheral once we are done with it. This is done by invoking the `close` method.

 Unless it is a special case, you should `open` all your peripherals inside `onCreate` and `close` them inside `onDestroy`.

In general, you will want to open the peripherals inside `onCreate` and close them inside `onDestroy`, so you can use them anytime along the lifecycle of the activity. You can, however, open, interact, and close them in a single block of code. It is less efficient if you are going to access them several times, but in some cases—such as single use—it could be preferred since the peripheral is held by the activity for just the minimum time needed.

There is also the concept of `AutoCloseable`, which is an interface that all the peripherals implement. It helps the system identify components that should be closed and it does so whenever the activity is destroyed.

 `AutoCloseable` peripherals should close automatically whenever they are deallocated, but it is still recommended to call `close` on all peripherals inside `onDestroy` to enforce clarity and consistency.

User space drivers

The architecture of Android Things is composed of two stacks: one managed by Google containing the kernel, hardware libraries, and the Android Things framework, and one open to developers containing user drivers and apps.

User drivers are components registered from within apps that extend existing Android framework services. They allow any application to inject hardware events into the framework that other apps can process using the standard Android APIs.

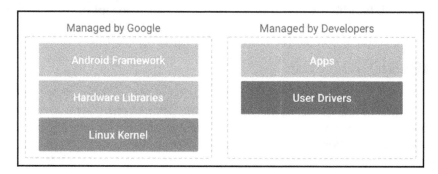

We will use this type of driver when we interact with the temperature and pressure sensor. That specific type of user driver uses sensor fusion to work that way. We will see how buttons can also be integrated as input drivers.

However, in general we can just use a simple driver that handles the communication with the peripheral we want to integrate with. These are just device drivers, and they do not need to be integrated into the platform.

And we can start with the equivalent of the `Hello World` for Internet of things: blinking an LED.

LEDs

If you are familiar with how Arduino code is structured, you may be expecting to have a `setup` and a `loop` methods. Android Things can work like that too. We will start with our methods named using that convention to provide a common ground if you have experience with Arduino. However, Android Things allows us to write code in a much more structured way. In this section we will start with something close to the Arduino way, explain its drawbacks, and show better alternatives, showing the different ways to blink an LED.

Blinking an LED is the IoT equivalent of `Hello World`.

Let's get into blinking an LED the Arduino way.

The Arduino way

The simplest way is to have a setup where we initialize the LED, followed by a loop that toggles the state and then waits.

The code for that looks like this:

```kotlin
class MainActivity : Activity() {

    private lateinit var led: Gpio

    override fun onCreate(savedInstanceState: Bundle?) {
        super.onCreate(savedInstanceState)
        setup()
        while (true) {
            loop()
        }
    }

    private fun setup() {
        led = RainbowHat.openLedRed()
    }

    private fun loop() {
        led.value = !led.value
        Thread.sleep(1000)
    }

    override fun onDestroy() {
        super.onDestroy()
        led.close()
    }
}
```

If you run this `Activity`, you'll see the red LED of the Rainbow HAT blinking. So, let's look at the different areas to see what is happening.

Inside `onCreate` we call `setup` and then make an infinite loop that simply calls the `loop` method.

Inside `setup` we open the LED using the utility method `RainbowHat.openLedRed()`. This abstraction allows us to open it without needing to know anything about where it is connected, but it still gives us the flow of opening and closing the components to respect the peripheral lifecycle. We store a reference to the object in a class variable so we can access it later.

Note that the `led` variable is declared as `lateinit`. That allows us to initialize it inside `onCreate` without requiring to declare it as nullable, which makes the rest of the code more readable.

The `led` variable is of type `Gpio`. We will look at GPIO in detail in `Chapter 3`, *GPIO - Digital Input/Output*. For now, what you need to know is that we can set the value to `true` or `false` and that will make the LED turn on and off.

Then the `loop` method toggles the LED value, sleeps for 1,000 milliseconds, and repeats.

Inside `onDestroy` we call the `close` method of the `led` variable, but since there is an infinite loop, this code never executes. However, since the `Gpio` class implements `AutoCloseable`, the LED is released.

Having an infinite loop inside `onCreate` is a very bad practice.

Having an infinite loop inside `onCreate` is a very bad practice. The `Activity` never completes the proper lifecycle—`onCreate` never finishes so `onResume` is never called—and that infinite loop is blocking the UI thread. So let's look at better approaches. We'll choose to use one or another depending on the circumstances. Don't use this approach; it is presented only as an example to show why it is a bad practice.

Threads

The obvious first step is to move the infinite loop outside of the UI thread. Since Android Things is based on Android, we have some classes that we can use to improve this, so let's use a `Thread` to do it. The updated code looks like this:

```
class BlinkThreadActivity : Activity() {

    private lateinit var led: Gpio

    private val thread = Thread {
        while(true) {
            loop()
        }
    }

    override fun onCreate(savedInstanceState: Bundle?) {
        super.onCreate(savedInstanceState)
        setup()
        thread.start()
    }

    private fun setup() {
        led = RainbowHat.openLedGreen()
    }

    private fun loop() {
        led.value = !led.value
        Thread.sleep(1000)
    }

    override fun onDestroy() {
        super.onDestroy()
        thread.interrupt()
        led.close()
    }
}
```

The overall concept is very similar; we still have a `setup` method that is executed at the beginning and a `loop` method that gets executed endlessly, but now inside `onCreate` we just create a thread that has the infinite loop and start it. This time we can interrupt the thread as part of `onDestroy`. So, as a first improvement, the UI thread is not blocked.

The other improvement is that `onCreate` does finish, so we do not tamper with the activity lifecycle.

Note that there is no need to modify the LED value on the UI thread. This is not UI in the Android sense, it is a communication done via a protocol to a wire; it is a different paradigm. It acts as a user interface, but since it is not composed of views, it is not rendered and does not require the UI Thread.

Interactions with peripherals do not need to be done on the UI thread.

One of the most powerful features of Kotlin is the integration with lambdas, and we can see it here in the definition of the thread. It gets rid of all the boilerplate code and it is a lot easier to read than the Java equivalent.

Threads require a significant amount of resources, and while our development boards are quite powerful, it is good to use simpler solutions. Another very interesting feature of Kotlin is coroutines, which we will be exploring next.

Coroutines

Coroutines are designed to provide a way to avoid blocking a thread when we need to perform an expensive operation and replace it with a cheaper and more controllable operation: `suspension` of a coroutine.

Coroutines simplify asynchronous programming. The logic of the program can be expressed sequentially in a coroutine and the underlying library will figure out the asynchrony for us. The code remains as simple as if it was written for sequential execution.

To learn more about coroutines, visit: `https://kotlinlang.org/docs/reference/coroutines.html`.

And, as we already mentioned, coroutines are cheaper than threads.

Since coroutines are not part of the basic core package of Kotlin yet, we need to add a line to our dependencies on the `build.gradle` to be able to use them:

```
dependencies {
    [...]
    implementation "org.jetbrains.kotlinx:kotlinx-coroutines-core:0.22.5"
}
```

Coroutines are complex and provide many interesting features, especially for syncing the executions of asynchronous code. We are going to use them in the most basic way: to simplify the loop, make it more readable than with a `Thread`, and more efficient on runtime.

The code for blinking an LED using a Kotlin coroutine is as follows:

```kotlin
class BlinkCoroutineActivity : Activity() {

    private lateinit var led: Gpio
    private lateinit var job: Job

    override fun onCreate(savedInstanceState: Bundle?) {
        super.onCreate(savedInstanceState)
        setup()
        job = launch {
            while(isActive) {
                loop()
            }
        }
    }

    private fun setup() {
        led = RainbowHat.openLedGreen()
    }

    private suspend fun loop() {
        led.value = !led.value
        delay(1000)
    }

    override fun onDestroy() {
        super.onDestroy()
        job.cancel()
        led.close()
    }
}
```

We can see several differences. The first one is the use of the `launch` keyword. It basically creates a coroutine and executes the lambda on it. It returns a variable that is the job itself.

We can also see that the `loop` method now includes the `suspend` modifier. This indicates that the method can be suspended. Again, we are just scratching the surface of coroutines.

Another difference is that we use `delay` instead of `Thread.sleep`, which is a more efficient way to handle waiting times.

Finally, instead of interrupting the thread, we just `cancel` the job.

But those methods are related to active waiting. We can and should do better. If we think about this problem the Android way, the tool we would use is either a `Handler` or a `Timer`, so let's explore both.

Using Handler and Runnable

From this example onward, we will be removing the `setup` and `loop` methods. We will include the code into the activity `onCreate` to keep it simpler. These methods were initially there to help when comparing with how it works on Arduino.

Now, the default way to reschedule events without waiting on Android is to use a `Handler`. This class allows us to post a `Runnable` object with an optional delay. This technique removes the need for waiting and/or sleeping.

Blinking an LED using a `handler` looks like this:

```kotlin
class BlinkHandlerActivity : Activity() {

    private lateinit var led: Gpio
    private val handler = Handler()

    private val ledRunnable = object: Runnable {
        override fun run() {
            led.value = !led.value
            handler.postDelayed(this, 1000)
        }
    }

    override fun onCreate(savedInstanceState: Bundle?) {
        super.onCreate(savedInstanceState)
        led = RainbowHat.openLedGreen()
        handler.post(ledRunnable)
    }

    override fun onDestroy() {
        super.onDestroy()
        handler.removeCallbacks(ledRunnable)
        led.close()
    }
}
```

We create a `Runnable` value that toggles the LED and then reschedules itself in the future by calling `postDelayed` on the handler.

We also schedule the runnable for the first time as part of onCreate, which will trigger the first execution.

Inside onDestroy, we just need to call removeCallbacks passing the ledRunnable object to prevent it from being called again, and, by doing so, breaking the infinite recursive call.

Doing it this way guarantees the time between executions. In this case the code inside ledRunnable is simple and will take very little time to execute, so it won't be noticeable. This is not always the case. While it is interesting to keep the time between executions constant, we may want to start each execution at a specific moment, ensuring that the distance between the start of each execution is constant. To do that, we use timers.

Using Timer and Timertask

When we use Timer and TimerTask, we are enforcing that each iteration is started at a proper interval, regardless of the time it takes to execute. Keep in mind that if the execution time is longer than the interval, the code will be executed twice in parallel and race conditions may appear.

 Timer and TimerTask guarantee that each execution starts at the proper interval.

Let's see how to blink an LED using Timer and TimerTask:

```kotlin
class BlinkTimerActivity : Activity() {

    private lateinit var led: Gpio
    private val timer = Timer()

    override fun onCreate(savedInstanceState: Bundle?) {
        super.onCreate(savedInstanceState)
        led = RainbowHat.openLedGreen()
        timer.schedule ( timerTask {
            led.value = !led.value
        }, 0, 1000 )
    }

    override fun onDestroy() {
        super.onDestroy()
        timer.cancel()
        timer.purge()
        led.close()
```

```
        }
    }
```

In this example, we create a `timer` value of type `Timer` and we schedule a `TimerTask` to be executed every 1,000 milliseconds. The code of the task is simply to toggle the LED.

Then, as part of `onDestroy`, we `cancel` and `purge` the timer, to stop the rescheduling and to remove any scheduled tasks.

In this case we guarantee that the call will start every second. Take a look at the following diagram to see the difference between a handler and a timer:

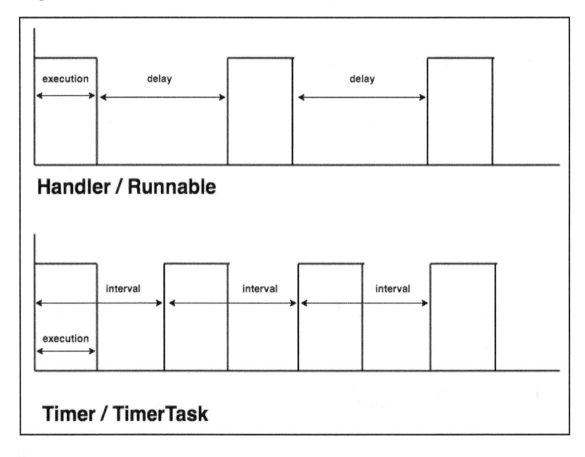

Buttons

The next component we are going to work with are capacitive buttons. The Rainbow HAT has three of them, labeled A, B, and C. Buttons can be handled in two different ways, using `Buttons` and using `InputDrivers`.

While `Buttons` are similar in behavior to the ones we place on a traditional UI, `InputDrivers` are similar to HID keyboards; they allow us to integrate with the operating system and send key press events, letting us handle our button presses as standard keys.

We will look at both of them with the same example: we will turn the red LED on when button A is pressed and off when it is not.

Let's start with the button driver.

Button driver

The simplest way to interact with a button is to use the button driver. This will allow us to interact with the physical button in a way similar to the traditional `OnClickListener` listener from `View`.

As usual, let's begin by looking at the code:

```kotlin
class ButtonDriverActivity : Activity() {

    private lateinit var redLed: Gpio
    private lateinit var buttonA: Button

    override fun onCreate(savedInstanceState: Bundle?) {
        super.onCreate(savedInstanceState)
        redLed = RainbowHat.openLedRed()
        buttonA = RainbowHat.openButtonA()
        buttonA.setOnButtonEventListener {
            button: Button, state: Boolean ->
            redLed.value = state
        }
    }

    override fun onDestroy() {
        super.onDestroy()
        redLed.close()
        buttonA.close()
    }
}
```

In this case we use the utility method `RainbowHat.openButtonA()` to create a `Button` object linked to the A button and open it.

Then we just set an `OnButtonEventListener`, which receives the `Button` and the state. The reason behind having the `Button` as a parameter is the same as why `OnClickListener` receives the `View`; we could use a listener for several items and we need to determine which one generated the event.

Inside the lambda, we simply set the `redLed` value to the current state of the button.

Lastly, as always, we close the button and the LED inside `onDestroy`.

 With buttons, we have event-based programming.

Note that we are doing event based-programming. Although this is the standard way to do things in Android, it is not that easy to do in Arduino. This is another advantage of Android Things.

Although the class name is `Button`, this class is not the `Button` from the Android UI framework. Make sure you have the correct import (sometimes auto-import takes the one from `android.view`).

 Make sure your button is imported from `com.google.android.things.contrib.driver.button.Button`.

There is a method of `Button` that is worth mentioning: `setDebouceDelay`.

Debouncing

When we are using a physical button, due to mechanical and physical issues, it is common that the parts make and lose contact several times when being pressed, generating spurious signals. This is called signal bouncing, and we need a way to filter those undesired events. That is what is called debouncing.

 Due to mechanical and physical issues it is common that buttons generate spurious signals when being pressed.

Debouncing can be done either by hardware or software.

`Button` has some debouncing code. It enforces that the signal must remain stable before generating an event. That stable threshold is set to 100 milliseconds by default. You can modify it, if needed, using `setDebouceDelay`.

 You can disable software debouncing by setting the debounce delay to 0.

If your buttons have a hardware debounce system, you may want to disable software debouncing. You can do that setting the value to zero. In that case, it disables debounce and triggers events on all edges immediately.

Button input drivers

The second approach to handle buttons is `ButtonInputDriver`. As we already mentioned, they are a type of user space driver and they integrate tightly with the operating system. Because of the level of integration, input drivers require the `MANAGE_INPUT_DRIVERS` permission, which we need to add to the manifest:

```
<uses-permission
android:name="com.google.android.things.permission.MANAGE_INPUT_DRIVERS" />
```

The lifecycle of a `ButtonInputDriver` is a bit different from the one of a peripheral. They register and unregister instead of open and close, but other than that (and the internals of it) the lifecycle is done in the same places. Let's see how it works:

```
class ButtonInputDriverActivity : Activity() {

    private lateinit var redLed: Gpio
    private lateinit var buttonInputDriverA: ButtonInputDriver

    override fun onCreate(savedInstanceState: Bundle?) {
        super.onCreate(savedInstanceState)

        redLed = RainbowHat.openLedRed()
        buttonInputDriverA =
        RainbowHat.createButtonAInputDriver(KeyEvent.KEYCODE_A)
        buttonInputDriverA.register()
    }

    override fun onDestroy() {
```

```
            super.onDestroy()
            redLed.close()
            buttonInputDriverA.unregister()
        }
    }
```

Again, the Rainbow HAT driver provides us with a wrapper that simplifies the creation of a `ButtonInputDriver` and abstracts us from the hardware. We just need to call `RainbowHat.createButtonAInputDriver(KeyEvent.KEYCODE_A)` to create it. The parameter to this method is the key code we want to associate with the button.

Again, a key difference is that we now *register* the driver with the system as part of `onCreate` instead of just opening the peripheral. Similarly, we need to `unregister` the driver inside `onDestroy`.

Once the driver is registered, we can handle the key events generated by it overriding `onKeyDown` and `onKeyUp`:

```
    override fun onKeyDown(keyCode: Int, event: KeyEvent?): Boolean {
        if (keyCode == KEYCODE_A) {
            redLed.value = true
            return true
        }
        else {
            return super.onKeyDown(keyCode, event)
        }
    }

    override fun onKeyUp(keyCode: Int, event: KeyEvent?): Boolean {
        if (keyCode == KEYCODE_A) {
            redLed.value = false
            return true
        }
        else {
            return super.onKeyUp(keyCode, event)
        }
    }
```

This is exactly the same as you would do for a real key press on an Android device.

As you can see, `Button` and `ButtonInputDriver` are quite different, and it will depend on your particular case to decide which one makes more sense or is better suited to deal with your project.

There is also a third -lower level- way to access buttons that also applies to sensors: registering GPIO callbacks. We will explore that in the next chapter when we dig deeper into GPIO.

 As an exercise, you can extend both examples linking each of the buttons with the LED that is placed over them.

Piezo buzzer

Let's make some noise. The buzzer also has its own driver. We can open it using the `RainbowHat.openPiezo()` method and it returns an object of type `Speaker`.

The `Speaker` class has two methods: `play`, which receives the frequency to play, and `stop`, which -unsurprisingly- makes the speaker stop.

To follow up from the previous section, let's build a three-tone piano where the buttons A, B, and C will play the frequencies of 1,000 Hz, 3,000 Hz, and 5,000 Hz, respectively. The buzzer will start sounding when the button is pressed and stop when the button is released.

For this example, we'll use button drivers for simplicity. The initialization and cleanup for `PianoActivity` looks like this:

```kotlin
class PianoActivity : Activity() {

    private lateinit var buttonA: ButtonInputDriver
    private lateinit var buttonB: ButtonInputDriver
    private lateinit var buttonC: ButtonInputDriver

    private lateinit var buzzer: Speaker

    private val frequencies = hashMapOf(
            KEYCODE_A to 1000.0,
            KEYCODE_B to 3000.0,
            KEYCODE_C to 5000.0)

    override fun onCreate(savedInstanceState: Bundle?) {
        super.onCreate(savedInstanceState)

        buzzer = RainbowHat.openPiezo()

        buttonA = RainbowHat.createButtonAInputDriver(KEYCODE_A)
        buttonB = RainbowHat.createButtonBInputDriver(KEYCODE_B)
```

```
        buttonC = RainbowHat.createButtonCInputDriver(KEYCODE_C)

        buttonA.register()
        buttonB.register()
        buttonC.register()
    }

    override fun onDestroy() {
        super.onDestroy()
        buzzer.close()
        buttonA.unregister()
        buttonB.unregister()
        buttonC.unregister()
    }

    [...]
}
```

We can see again the standard lifetime of the peripherals, being opened inside onCreate and closed inside onDestroy. As mentioned before, we used the utility wrapper method to open the piezo buzzer.

The other interesting part of initialization is the way we store the frequencies to play. We will have a Map that uses the key codes as keys and has the frequencies as values. What is interesting here is how easy it is to read the initialization of such variable in Kotlin.

Now, for the interesting part, let's look into onKeyDown and onKeyUp:

```
override fun onKeyDown(keyCode: Int, event: KeyEvent?): Boolean {
    val freqToPlay = frequencies.get(keyCode)
    if (freqToPlay != null) {
        buzzer.play(freqToPlay)
        return true
    }
    else {
        return false
    }
}

override fun onKeyUp(keyCode: Int, event: KeyEvent?): Boolean {
    buzzer.stop()
    return true
}
```

Besides the code of the button handling and finding the right frequency to play, the only code related to the piezo buzzer are the calls to play and stop.

Note that when we get the frequency from the map, it can be `null`. Kotlin enforces `null` validation because a key that is not referenced on the map may get pressed. We don't want our application to crash, so if there is no value for the key, we just pass the event up the chain to the parent class. This is a good example of how Kotlin handles nullables and why it is one of its best features.

Alphanumeric display (Ht16k33)

The most prominent part of the Rainbow HAT is the four-digit, 15-segment alphanumeric display. It takes most of its surface and it is also very handy, since it can be used to display text.

`contrib-drivers` includes a class named `AlphanumericDisplay`, which extends from `Ht16k33` (the controller chip) and it adds a few string manipulation utilities.

The simplest way to use the display is just five lines of code:

```
val alphanumericDisplay = RainbowHat.openDisplay()
alphanumericDisplay.setBrightness(Ht16k33.HT16K33_BRIGHTNESS_MAX)
alphanumericDisplay.setEnabled(true)
alphanumericDisplay.display("AHOY")
alphanumericDisplay.close()
```

As usual, we use the `RainbowHat` utility method to open the peripheral (in this case, `RainbowHat.openDisplay`), then we set the brightness (in our example, to the maximum value), and we also set it to `enabled`. Setting the display to `enabled` is very important, because otherwise it will not work.

 If you forget to call `setEnable(true)`, the alphanumeric display will not show anything.

Note that we are working with hardware, and it persists the state outside of our application. So, if you previously run another application that enabled it, it will stay enabled when you open it, even if you forget to call `setEnable` in your app, but it won't work once you reboot the developer kit. That is a very hard to debug situation.

Then we just invoke the `display` method with the text to be displayed. There are several variations of display that receive different types as parameters for versatility.

Finally, as with any other peripheral, we need to call `close` when we are done working with it. We will be using the alphanumeric display in several examples.

Temperature and pressure sensor (Bmx280)

The temperature and pressure sensor on the Rainbow HAT uses the BMP280 chip. A similar component—BME280—has an extra humidity sensor and the driver is designed to work with both chips (hence, the X in the name Bmx280) since both have the same internal protocol. Again, remember that the one in the Rainbow HAT does not include a humidity sensor.

Android Things offers two ways to read data from sensors. The first one is to proactively read a value from the component itself, and the second one is to configure a `SensorDriver` that will deliver readings via a listener whenever the values change. This is meant to use the same framework as the sensors on a phone (namely gyroscope, magnetometer, and accelerometer).

Querying the component directly is simpler; it gives all the control and also all the responsibility to us. On the other hand, sensor drivers rely more on the system itself and are more integrated. It is similar to the differences between using `Button` and `InputDriver` that we explored earlier in the chapter.

Let's start with the simpler approach.

Direct read

For this example, we are going to read the temperature and then use the alphanumeric display to show the value we obtained from the sensor. We will also use a handler to query the driver for new values. The use of the handler is very similar to what we did for the LEDs earlier except that, in this case, we will use it to provide continuous readings using the `post` method instead of `postDelayed`.

```
class TemperatureDisplayActivity: Activity() {

    private val handler = Handler()

    private lateinit var sensor: Bmx280
    private lateinit var alphanumericDisplay: AlphanumericDisplay

    val displayTemperatureRunnable = object: Runnable {
        override fun run() {
```

```kotlin
            val temperature = sensor.readTemperature().toDouble()
            // Display the temperature on the alphanumeric display
            alphanumericDisplay.display(temperature)
            handler.post(this)
        }
    }

    override fun onCreate(savedInstanceState: Bundle?) {
        super.onCreate(savedInstanceState)

        sensor = RainbowHat.openSensor()
        sensor.temperatureOversampling = Bmx280.OVERSAMPLING_1X
        sensor.setMode(Bmx280.MODE_NORMAL)

        alphanumericDisplay = RainbowHat.openDisplay()
        alphanumericDisplay.setBrightness(Ht16k33.HT16K33_BRIGHTNESS_MAX)
        alphanumericDisplay.setEnabled(true)

        handler.post(displayTemperatureRunnable)
    }

    override fun onDestroy() {
        super.onDestroy()
        handler.removeCallbacks(displayTemperatureRunnable)
        sensor.close()
        alphanumericDisplay.close()
    }
}
```

Our activity has two member variables, which we initialize inside onCreate: sensor, of type Bmx280 and alphanumericDisplay of type AlphanumericDisplay.

There is also a Runnable object, which we use in combination with the handler to check the temperature continuously.

Inside onCreate we open and initialize the sensor using RainbowHat.openSensor() and the alphanumeric display in the same way we did before.

Note that we set the value for temperature oversampling and the mode to normal. Oversampling is required for the component to start checking the temperature. If we skip it, we won't be able to read the temperature at all. Setting the mode to normal makes sure that the sensor is not in sleep mode. When the sensor is in sleep mode the readings are static. That mode is useful to save power when we are not actively reading from it.

Finally, we do the standard cleanup inside `onDestroy`, removing callbacks from the handler and closing the sensor and the alphanumeric display.

Unfortunately, the temperature sensor sits right on top of the CPU on the Raspberry Pi, and quite close to the LCD display, so you will be seeing higher than expected values, as well as a steady increase of the temperature. That is not a fault on your code; it is just a flaw in the design of the HAT. If you are planning to use these readings for anything serious, consider getting an external sensor.

The temperature sensor is positioned right on top of the CPU on the Raspberry Pi, and quite close to the LCD display. Don't use the temperature readings from the Rainbow HAT for anything serious.

If instead of the temperature we want to read the pressure, the code is almost identical:

```
class PressureDisplayActivity: Activity() {

    private val handler = Handler()

    private lateinit var sensor: Bmx280
    private lateinit var alphanumericDisplay: AlphanumericDisplay

    val displayPressureRunnable = object: Runnable {
        override fun run() {
            val pressure = sensor.readPressure().toDouble()
            alphanumericDisplay.display(pressure)
            handler.post(this)
        }
    }

    override fun onCreate(savedInstanceState: Bundle?) {
        super.onCreate(savedInstanceState)

        sensor = RainbowHat.openSensor()
        sensor.temperatureOversampling = Bmx280.OVERSAMPLING_1X
        sensor.pressureOversampling = Bmx280.OVERSAMPLING_1X

        alphanumericDisplay = RainbowHat.openDisplay()
        alphanumericDisplay.setBrightness(Ht16k33.HT16K33_BRIGHTNESS_MAX)
        alphanumericDisplay.setEnabled(true)

        handler.post(displayPressureRunnable)
    }

    override fun onDestroy() {
```

```
        super.onDestroy()
        handler.removeCallbacks(displayPressureRunnable)
        sensor.close()
        alphanumericDisplay.close()
    }
}
```

As you can see, the only differences are that we replace `temperature` with `pressure` inside the `runnable` object and that we configure oversampling for both `temperature` and `pressure`. This is needed because the algorithm that calculates value rectification for pressure requires the temperature as input.

 To be able to read `pressure`, we need to configure oversampling for `temperature` and `pressure`.

Now that we have seen how to read from the sensor, let's take a look at how to do it using `SensorDriver`.

Continuous updates with sensor driver

Similar to what happened with the button input driver, when we use classes that integrate deeply with the operating system, we need special permissions. Any time we want to create new sensor drivers we need to add the `MANAGE_SENSOR_DRIVERS` permission to the manifest:

```
<uses-permission
android:name="com.google.android.things.permission.MANAGE_SENSOR_DRIVERS"
/>
```

Using a sensor driver is more complicated than reading directly from the peripheral. Let's split the code into two parts: the initialization of the sensor driver, and the callbacks and listeners. Initialization looks similar to what we did for button input drivers:

```
class TemperatureSensorDriverActivity : Activity() {

    lateinit var sensorManager : SensorManager
    lateinit var sensorDriver : Bmx280SensorDriver

    override fun onCreate(savedInstanceState: Bundle?) {
        sensorManager = getSystemService(Context.SENSOR_SERVICE) as
SensorManager
        sensorManager.registerDynamicSensorCallback(sensorCallback)
```

```
            super.onCreate(savedInstanceState)
            sensorDriver = RainbowHat.createSensorDriver()
            sensorDriver.registerTemperatureSensor();
    }


    [.....]
}
```

Initialization of the sensor driver comes in two steps. First we register a dynamic sensor callback with the SensorManager. This listener will be invoked when a new sensor is connected. When this happens, we register the temperature sensor into the system. This step will add the sensor to the list of sensors available.

Let's take a look into the callbacks now:

```
val sensorCallback = object : SensorManager.DynamicSensorCallback() {
    override fun onDynamicSensorConnected(sensor: Sensor?) {
        if (sensor?.type == Sensor.TYPE_AMBIENT_TEMPERATURE) {
            sensorManager.registerListener(
                temperatureSensorListener,
                sensor,
                SensorManager.SENSOR_DELAY_NORMAL)
        }
    }
}

val temperatureSensorListener = object : SensorEventListener {
    override fun onSensorChanged(event: SensorEvent) {
        Log.i(TAG, "Temperature changed: " + event.values[0])
    }

    override fun onAccuracyChanged(sensor: Sensor, accuracy: Int) {
        Log.i(TAG, "accuracy changed: $accuracy")
    }
}
```

The dynamic sensor callback receives onDynamicSensorConnected when the new sensor is connected. Then, inside the method, we check if the sensor is of type TYPE_AMBIENT_TEMPERATURE and, if so, we register a SensorEventListener to get updates from this sensor.

The registration includes the listener, the sensor, and the sampling rate.

The last part is the SensorEventListener itself. We only care about the calls to onSensorChanged, which will return us a list of values with the sensor readings. In our case, this list has just one value.

`SensorEvent` contains an array of values. This makes sense because `SensorEventListener` is used by all sensors, including complex ones such as accelerometers, magnetometers, and so on.

Finally, let's look at the cleanup:

```
override fun onDestroy() {
    super.onDestroy()
    sensorManager.unregisterListener(temperatureSensorListener)
    sensorDriver.unregisterTemperatureSensor()
    sensorDriver.close()
}
```

In a similar way as the one to initialize the sensor, the cleanup inside `onDestroy` consists of the unregistration of the sensor listener and the unregistration of the sensor with the system. The dynamic sensor callback is irrelevant at this time, so we can choose not to unregister, but you can do that as well to have everything tidied up.

Adding pressure driver

As an extended example, let's look at how we can handle pressure with a sensor driver.

Inside the dynamic sensor callback, we add another check for sensors of the type `TYPE_PRESSURE` and we register another `SensorEventListener`. The listener is the same as the one for temperature, but this one is registered only to listen for pressure.

```
val sensorCallback = object : SensorManager.DynamicSensorCallback() {
    override fun onDynamicSensorConnected(sensor: Sensor?) {
        if (sensor?.type == Sensor.TYPE_AMBIENT_TEMPERATURE) {
            registerTemperatureListener(sensor)
        }
        if (sensor?.type == Sensor.TYPE_PRESSURE) {
            registerPressureListener(sensor)
        }
    }
}

private fun registerPressureListener(sensor: Sensor) {
    sensorManager.registerListener(
        pressureSensorListener,
        sensor,
        SensorManager.SENSOR_DELAY_NORMAL)
}
```

```
val pressureSensorListener = object : SensorEventListener {
    override fun onSensorChanged(event: SensorEvent) {
        Log.i(TAG, "pressure changed: " + event.values[0])
    }

    override fun onAccuracyChanged(sensor: Sensor, accuracy: Int) {
        Log.i(TAG, "accuracy changed: $accuracy")
    }
}
```

And, finally, we add the call to `registerPressureSensor` into `onCreate` and the equivalent `unregisterPressureSensor` into `onDestroy`, as well as the one to unregister the listener:

```
override fun onCreate(savedInstanceState: Bundle?) {
    super.onCreate(savedInstanceState)
    sensorDriver = RainbowHat.createSensorDriver()
    sensorDriver.registerTemperatureSensor();
    sensorDriver.registerPressureSensor()

    sensorManager = getSystemService(Context.SENSOR_SERVICE) as
SensorManager
    sensorManager.registerDynamicSensorCallback(sensorCallback)
}

override fun onDestroy() {
    super.onDestroy()
    sensorManager.unregisterListener(temperatureSensorListener)
    sensorManager.unregisterListener(pressureSensorListener)
    sensorDriver.unregisterTemperatureSensor()
    sensorDriver.unregisterPressureSensor()
    sensorDriver.close()
}
```

With that, we can read temperature and pressure using the listeners and callbacks from the Android framework.

LED strip (Apa102)

For the last part, we have the RGB LED strip, which is the one that gives the Rainbow HAT its name. It has seven RGB LEDs that we can set to any combination of colors.

For this example, we will just use the color red, because we are going to build a moving LED like that of KITT from Knight Rider or, if you prefer, from a Cylon eye. Having such pop culture references at hand, our example could not be any other one.

For this example, we will use a timer and keep track of the current position that is lit up, and also check edge conditions to bounce and change direction.

Let's start with the initialization and cleanup code:

```
class KnightRiderSimpleActivity : Activity() {
    private val timer: Timer = Timer()
    private var goingUp = true
    private var currentPos = 0
    private val interval: Long = 100

    val colors = IntArray(RainbowHat.LEDSTRIP_LENGTH)
    private lateinit var ledStrip: Apa102

    override fun onCreate(savedInstanceState: Bundle?) {
        super.onCreate(savedInstanceState)

        ledStrip = RainbowHat.openLedStrip()
        ledStrip.brightness = Apa102.MAX_BRIGHTNESS
        ledStrip.direction = Apa102.Direction.NORMAL

        timer.schedule(timerTask {
            knightRider()
        }, 0, interval)
    }

    override fun onDestroy() {
        super.onDestroy()
        timer.cancel()
        ledStrip.close()
    }

    [...]
}
```

We open the LED strip using the utility method from `RainbowHat`, in the same way as we have been doing for all peripherals, in this case with the `openLedStrip` method.

The initialization is completed by setting the brightness and the direction.

Note that the brightness is a single value for all the LEDs; the chip, however, supports independent brightness. It is a good exercise to read the driver code and modify it to support it, but if you feel lazy, there is a branch of the project, `AndroidThingsKnightRider`, that has it implemented.

The driver is generic for LED strips of any length. We use `RainbowHat.LEDSTRIP_LENGTH` to define the length of the one in the HAT.

Direction means in which order are we going to send the values. It is similar to little endian and big endian. Normal means the LED on the left is at position 0, and the one farthest right is at `LEDSTRIP_LENGTH-1`.

The driver allows only to set the brightness globally, but the peripheral does support independent brightness.

The last part in the initialization is the creation of the timer and the schedule of a timer task that calls the `knightRider` method.

The cleanup part inside `onDestroy` only needs to take care of closing the `ledStrip` variable and canceling the timer.

Note that we have created an array of integers of length `RainbowHat.LEDSTRIP_LENGTH`. That is what we will use to send the color information to the driver.

The function inside the timer task is where the action happens. It is composed of two parts: `updateCurrentPos` and `updateLedStrip`:

```
private fun knightRider() {
    updateCurrentPos()
    updateLedStrip()
}
```

Updating the position increases or decreases the current position based on the value of `goingUp`. It then toggles that value whenever it reaches the end of the LED strip:

```
private fun updateCurrentPos() {
    if (goingUp) {
        currentPos++
        if (currentPos == RainbowHat.LEDSTRIP_LENGTH - 1) {
            goingUp = false
        }
    } else {
        currentPos--
        if (currentPos == 0) {
```

```
                goingUp = true
            }
        }
    }
```

Finally, we update the values in the colors array. We set the current position to `Color.RED` and the other ones to `Color.TRANSPARENT`. Then we simply invoke `write` on the `ledStrip` passing the `colors` to send them to the peripheral. Note that the array must be of the correct length of the LED strip; otherwise, it will not work properly.

```
private fun updateLedStrip() {
    for (i in colors.indices) {
        if (i == currentPos) {
            colors[i] = Color.RED
        } else {
            colors[i] = Color.TRANSPARENT
        }
    }
    ledStrip.write(colors)
}
```

This is a very simple, yet satisfying example. Please play around with the colors array and use different colors for each LED. Although not as handy as the alphanumeric display, it is probably the most fun to play with of all the Rainbow HAT components.

Summary

In this chapter, we have learned about the Android Things architecture, and then worked with a lot of different hardware components: LEDs, buttons, buzzer, alphanumeric display, sensor reading, and LED strip, with lots of examples and coding guidelines.

Although all the components seemed identical, we have used four different communication protocols: GPIO, PWM, I2C, and SPI.

The Rainbow HAT meta driver has abstracted us from these differences, but as soon as we move away from it and into using other components, we will need to know which protocols they use to communicate and how to wire them, even if it is just to know which pin to connect them to.

If we flip the Rainbow HAT, we can see the specs of each of the components:

- **LED Red, Green, and Blue**: BCM6, BCM19, and BCM26
- **Buttons A, B, and C**: BCM21, BCM20, and BCM16
- **Piezo**: PWM1
- **Alphanumeric display (HT16K33)**: I2C1 0x70
- **Baro/Temp (BMP280)**: I2C1 0x77
- **Rainbow (APA102)**: SPI0.0

This indicates the pin where they are connected and also the protocol underneath (BCM is the name of the GPIO pins on a Raspberry Pi). Note that since this HAT was originally designed for Raspberry Pi, it has the pin name for that board. Pin names for Pico iMX7D are different and we will also see how to solve this problem in the next chapter.

In the following four chapters we will be looking at each one of those protocols in more detail, explain briefly how they work, see how we can access the ones in the HAT without the meta driver, and we'll also take a look at other components that use the same protocol that are common and interesting to use.

Let's move on to GPIO.

GPIO - Digital Input/Output

3

Now that we have seen all the hardware working with the Rainbow HAT in the previous chapter, we are going to learn about the first communication protocol: **GPIO** (**General-Purpose Input/Output**). We have used it already for LEDs as output and for buttons as input, and we will extend that to other outputs and inputs. Finally, we will look at a few components that use GPIO in a more generic way (DC motor controller, stepper motor, distance sensor, and LCD display).

GPIO is the simplest protocol to control devices; it just uses digital signals that can be on or off. In digital circuits, there are two possible values for what is considered a logic 1: 3.3v and 5v. The current Android Things developer kits use 3.3v. Note that the developer kits have some 5v Vcc pins, but that is used exclusively to power external circuits and has nothing to do with the value of GPIO.

GPIO in Android Things developer kits is 3.3v.

Note that the ports do not have predetermined usage, hence the general part of the name, and some components use them in a very particular way. We can configure a GPIO pin to be used for input or output at runtime, and even change that dynamically.

A very important aspect of GPIO is that it is used for signaling and it is not intended to be used for power. Making an LED blink is about as much power as you should drain from GPIO. Do not use GPIO to power anything!

Never use a GPIO pin to power an external circuit; it may damage your developer kit.

One last important note is that the Rainbow HAT does not expose any GPIO pins, so we will have to take it out to connect the extra hardware.

And remember that, when using any peripherals, you need to request the permission USE_PERIPHERAL_IO in the manifest.

```
<uses-permission
android:name="com.google.android.things.permission.USE_PERIPHERAL_IO" />
```

This chapter covers the following topics:

- Making the code work on any developer kit
- Using GPIO for output
- Using GPIO for input
- Other usages of GPIO

Let's get started with the generic way to address pins in a way that supports different boards.

Technical requirements

You will be required to have Android Studio and Android Things installed on a developer kit. You also will require many hardware components to effectively perform the tasks given in this chapter. The components are very interesting to have, just to see them working, but the Rainbow HAT is particularly important. We go into details about the developer kits and how to pick the right one, as a part of Chapter 1, *Introducing Android Things*. Finally, to use the Git repository of this book, you need to install Git.

The code files of this chapter can be found on GitHub:
https://github.com/PacktPublishing/Android-Things-Quick-Start-Guide.

Check out the following video to see the code in action:

http://bit.ly/2PnjkoX.

Making the code work on any developer kit

As we mentioned before, the pin names of the Raspberry Pi and the iMX7D are different. You can take a look at the pinout diagrams in the Appendix, and I suggest that you print them out to have them at hand.

Have the pinout diagram of your developer kit at hand.

When we flipped the Rainbow HAT, we could see that the red LED was wired to BCM6, which is pin 31 of the board. While the pin position is the same, that pin on the iMX7D is called GPIO2_IO02.

However, if we look at the source code of the Rainbow HAT driver, which is in Java, we see that it is accessed using a function.

```
public static Gpio openLedRed() throws IOException {
    return openLed(BOARD.getLedR());
}
```

This is a concept that is used frequently on examples when you want them to work on both devices. At the very minimum you should use a constant for the pin, so if someone has to change it, it is easy to find and update, but having an abstraction is better. This utility is normally called BoardDefaults.

 We can identify the board using Build.DEVICE.

The value of Build.DEVICE will tell us which development board we are running in: rpi3 or imx7d_pico; then we can return the values of the pins based on that.

The Rainbow HAT driver uses an interface and then two implementations, one per board. That makes sense for the amount of pins it is using, but if our project uses just a few pins, it is simpler to wrap the logic into each method. Even more, when using Kotlin, we can get one step further and include that code into the get of the constant and use a when selector, like this:

```
object BoardDefaults {
    private const val DEVICE_RPI3 = "rpi3"
    private const val DEVICE_IMX7D_PICO = "imx7d_pico"

  val ledR: String
    get() = when (Build.DEVICE) {
        DEVICE_RPI3 -> "BCM6"
        DEVICE_IMX7D_PICO -> "GPIO2_IO02"
        else -> throw IllegalStateException("Unknown Build.DEVICE
```

```
${Build.DEVICE}")
  }
}
```

You can have this object as part of your class, but it is recommended to have it on its own file to separate hardware configuration from logic.

We will include a couple of examples of BoardDefaults throughout the chapter so you can get used to them, but ultimately they are kind of boilerplate code, so we will let them out for clarity.

Time to look at the usage of GPIO in detail.

Using GPIO for output

Digital output is used to control devices. LEDs are just an easy and visual way to check that things work, and we will start by revisiting them to set the concepts and then look at relays, which are one of the most versatile components we can use.

LEDs

In this section we will see how to handle LEDs directly, learn a bit more about PeripheralManager (the class you use to access peripherals), and understand how GPIO works for output.

So, let's begin by replacing the utility method from the Rainbow HAT meta driver with some code that accesses the pins directly.

Given that we are already using a Gpio object to handle the LEDs, the only part that we need to change is how to open it. We already looked at openLedRed() from RainbowHat:

```
public static Gpio openLedRed() throws IOException {
  return openLed(BOARD.getLedR());
}
```

We also looked at `BOARD.getLedR()` and how it gives us the pin name which is connected to the Rainbow HAT red LED according to the board. The other part is a generic function to open an LED. Let's look at that code:

```
public static Gpio openLed(String pin) throws IOException {
  PeripheralManager pioService = PeripheralManager.getInstance();
  Gpio ledGpio = pioService.openGpio(pin);
  ledGpio.setDirection(Gpio.DIRECTION_OUT_INITIALLY_LOW);
  return ledGpio;
}
```

Here we have the real deal. First we get an instance of `PeripheralManager` and then we use it to open a `Gpio` pin by name.

> `PeripheralManager`, as the name indicates, is the class we use to manage all peripherals.

Once we have the `Gpio` object, we call `setDirection`, passing `Gpio.DIRECTION_OUT_INITIALLY_LOW`. This step is very important. Since GPIO can be used for input or output, when we set the direction, we are indicating how we want to use that particular pin. Unless we specify the direction we want, the system cannot consider the pin open since it will not know how to handle it.

In this case we are setting it as output, but also indicating the initial state. As you may have guessed, `Gpio.DIRECTION_OUT_INITIALLY_HIGH` also exists.

If we now replace the code inside one of our examples of the previous chapter, it would look like this:

```
class BlinkTimerActivity : Activity() {
    override fun onCreate(savedInstanceState: Bundle?) {
        super.onCreate(savedInstanceState)
        led = PeripheralManager.getInstance().openGpio(BoardDefaults.ledR)
        led.setDirection(Gpio.DIRECTION_OUT_INITIALLY_LOW)
        timer.schedule(timerTask {
            led.value = !led.value
        }, 0, 1000)
    }

    [...]
}
```

The only difference with the previous chapter is that now we open it using `PeripheralManager` directly, since we were already using a `Gpio` reference to handle the LED.

`PeripheralManager` is one the most important classes of the Android Things library, it is the one we use to interact with all the peripherals. It was used behind the scenes on the Rainbow HAT driver, and it is usually done that way on drivers.

There is one last method we haven't discussed yet: `setActiveType`. Normally you would want to have `active` as high, but not always:

```
led.setActiveType(Gpio.ACTIVE_HIGH)
```

The `active` type maps the value of `true` or `false` to the value of `high` and `low`, as you can see in the following diagram

		Active type	
		HIGH	LOW
Value	True	1	0
	False	0	1

Just to double-check: when we set active as low, a value of false means that we do have signal on the pin.

Let's move on to something more interesting to do with our digital outputs: controlling anything with a relay.

Relays

Relays are very useful components; they allow us to automate pretty much any device:

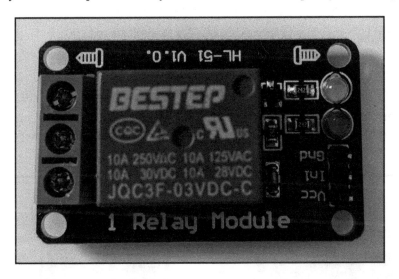

A relay is essentially a digital switch. This means that it works like a switch, but instead of a physical button, it uses a digital signal.

Relays have two sets of pins. One is connected to our development board, including Vcc, Ground, and a signal pin -which is the GPIO-. The other part is the one we automate -the switch- which consists of an input (typically the middle pin) and two outputs, one that is connected when the signal is low (typically labelled NC) and one that is connected when the signal is high. Quite often only two are used, but the third one is handy in certain cases.

 Relays allow us to automate pretty much anything.

Relays can work with AC as well as with DC, so you can use them to control something such as an electric popcorn maker (which uses AC) as well as a simple DC motor.

A very important concept for a relay is that it separates the signal from the actual current that powers the device. It enforces the fact that we should not use GPIO to power anything.

Make sure your relays are 3.3v, since that is the value of a GPIO on the Android Things developer kit. Note that most Arduino boards use 5v as GPIO value, so many relays you find online will not work with Android Things straight away. You will need to make a small circuit to make them work, but you don't want to need to do this.

Make sure your relays are activated at 3.3v.

The code to handle a relay module is the same as for lighting up the LED: just a simple GPIO configured for output. The wiring, however, is a bit different, as you can see in the following diagram:

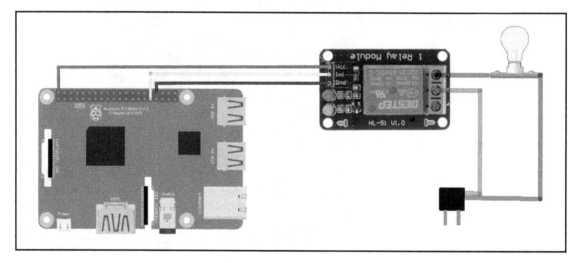

To be able to reuse the same code as for blinking the red LED, you just need to take the Rainbow HAT off and connect the relay to the pin 31 of the board (which is BCM6 or GPIO2_IO02, depending on your developer kit).

Note again that the current is coming from a main plug and the GPIO value is simply used as a signal to control the switch.

And, with that, you can blink a real bulb. We'll see a more interesting example with relays in the next section, when we talk about input.

Using GPIO for input

Now that we have seen how to use GPIO for output, let's move on to input. In this section we will look at accessing `Button` and `ButtonDriver` without using the meta driver of the Rainbow HAT. Both drivers are a layer of abstraction over GPIO, so our next step will be to learn how to work with GPIO directly. Finally, we will look at a few sensors that generate a GPIO signal and how to use them.

Buttons

Buttons are quite simple to deal with when using the driver or the input driver, but they can also be used to learn how to use GPIO output at a low level.

Let's get into the button driver.

Button driver

First, we will remove the Rainbow HAT meta driver from `dependencies` and add an entry for the button driver instead. This driver includes both the button and the button input drivers.

```
dependencies {
    [...]
    implementation 'com.google.android.things.contrib:driver-button:+'
}
```

In a similar way as we did for LEDs, let's look at the code of the wrapper function:

```
public static Button openButtonA() throws IOException {
    return openButton(BOARD.getButtonA());
}
```

And, as we saw for the LEDs, a `BOARD` object is used to provide the correct name of the pin depending on the development board and then call a generic function that opens a button. Let's see the code of the generic `openButton`:

```
public static final Button.LogicState BUTTON_LOGIC_STATE =
Button.LogicState.PRESSED_WHEN_LOW;

public static Button openButton(String pin) throws IOException {
    return new Button(pin, BUTTON_LOGIC_STATE);
}
```

Since we are using a driver, this step is very simple; the only thing the meta driver hides from us is the logic state of the buttons. Since they are capacitive, they are pressed when the GPIO is low (which is the reverse of normal press buttons). This is the equivalent of the active type we saw with inputs in the previous section.

So, if we just put everything together, we can just collapse it on a single line:

```
buttonA = Button(BoardDefaults.buttonA, Button.LogicState.PRESSED_WHEN_LOW)
```

And then, just for completeness, we add a new value to the `BoardDefaults` object for the pin where the button is connected:

```
val buttonA: String
    get() = when (Build.DEVICE) {
        DEVICE_RPI3 -> "BCM21"
        DEVICE_IMX7D_PICO -> "GPIO6_IO14"
        else -> throw IllegalStateException("Unknown Build.DEVICE
${Build.DEVICE}")
    }
```

Let's take the example of the previous chapter that used a button and an LED and see how it will look when we completely remove the abstraction for the Rainbow HAT:

```
class ButtonDriverActivity : Activity() {

    private lateinit var redLed: Gpio
    private lateinit var buttonA: Button

    override fun onCreate(savedInstanceState: Bundle?) {
        super.onCreate(savedInstanceState)

        redLed =
PeripheralManager.getInstance().openGpio(BoardDefaults.ledR)
        redLed.setDirection(Gpio.DIRECTION_OUT_INITIALLY_LOW)

        buttonA = Button(BoardDefaults.buttonA,
Button.LogicState.PRESSED_WHEN_LOW)
        buttonA.setOnButtonEventListener { button: Button, state: Boolean
->
            redLed.value = state
        }
    }

    override fun onDestroy() {
        super.onDestroy()
        redLed.close()
        buttonA.close()
```

```
        }
    }
```

There is not much of a difference, except that now we need to know more about our components, such as the logic state of the button and the pins where they are connected.

Note that the button driver hides the invocation to `PeripheralManager`, which we needed to use for the LED. It is still there, but enclosed inside the code of the driver. Taking care of opening the pins and interacting with `PeripheralManager` is a common pattern that most drivers follow.

Input drivers

The case of button input drivers is similar to that of button drivers. If we look again at the Rainbow HAT driver, we find out that we can make an input driver just with this line of code:

```
buttonDriverA = ButtonInputDriver(
    BoardDefaults.buttonA,
    Button.LogicState.PRESSED_WHEN_LOW,
    KEYCODE_A)
```

As previously, we need to specify the pin, -even if we do it via `BoardDefaults`- and the logic state. However, in this case, we also need to specify the key code that we want to bind it to.

We can go one step further and access GPIO for input without any intermediate drivers, similar to what we did with LEDs.

Controlling buttons directly with GPIO

If we look inside the code of the button driver, we can see how the GPIO pins are handled. The driver itself is little more than a wrapper that simplifies opening and provides software debouncing.

Let's see how to handle a button using GPIO directly:

```
val buttonA =
PeripheralManager.getInstance().openGpio(BoardDefaults.buttonA)
buttonA.setDirection(Gpio.DIRECTION_IN)
buttonA.setActiveType(Gpio.ACTIVE_LOW)
buttonA.setEdgeTriggerType(Gpio.EDGE_BOTH)
buttonA.registerGpioCallback {
    [...]
```

```
    true
}
```

As expected, the code opens the GPIO pin using `PeripheralManager`.

Then the direction is set to `Gpio.DIRECTION_IN` which configures it as an input, and the active type as `Gpio.ACTIVE_LOW` which is the translation of `LogicState.PRESSED_WHEN_LOW`. All of this is quite what we were expecting.

Something new here is that we set the edge trigger, which will define which events we want to be notified of. The possible values of edge trigger are as follows:

- `EDGE_NONE`: No events are generated. Note that this is the default value. If we do not define an edge type, the GPIO will not send any event.
- `EDGE_RISING`: Generates events on a transition from low to high (regardless of active type).
- `EDGE_FALLING`: Generates events on a transition from high to low (regardless of active type).
- `EDGE_BOTH`: Generates events on all state transitions.

For buttons it makes sense to use `EDGE_BOTH`, but other sensors, or other use cases, could require a different edge trigger.

Finally, we attach a `GpioCallback` to be executed when an event of the selected type happens.

Note that the last line of the lambda function for `onGpioEdge` is a return true statement. Returning `true` indicates that the listener should continue receiving events for each port state change.

But not all GPIO inputs are buttons; let's take a look at other sensors.

Other sensors

There are a few sensor types that provide digital signals. Some of them are digital by nature, such as presence and door sensors, and most commonly, they are analog sensors with an external circuit that implements a threshold, usually attached to a potentiometer, such as smoke/harmful gas detectors or raindrop detectors.

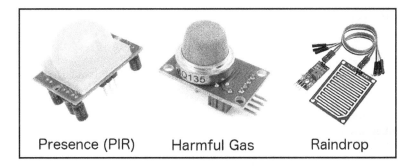

Several types of digital sensors

We will look at analog sensors in a later chapter. For now let's focus on sensors with a digital signal.

Let's look at a diagram that uses a presence sensor to turn a lamp on or off:

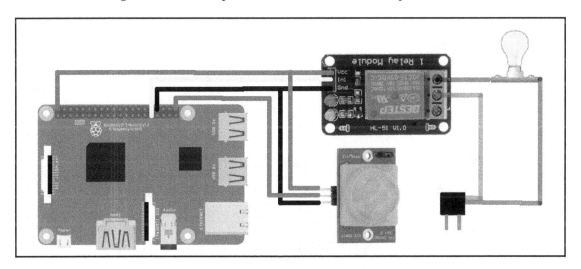

Replacing button A with a PIR sensor and the red LED with a bulb

Note that the wiring and components are completely different from what we have been using until now. However, let's look at the code:

```
class ButtonGpioActivity : Activity() {

    private lateinit var redLed: Gpio
    private lateinit var buttonA: Gpio

    override fun onCreate(savedInstanceState: Bundle?) {
```

```
        super.onCreate(savedInstanceState)

        redLed =
PeripheralManager.getInstance().openGpio(BoardDefaults.ledR)
        redLed.setDirection(Gpio.DIRECTION_OUT_INITIALLY_LOW)

        buttonA =
PeripheralManager.getInstance().openGpio(BoardDefaults.buttonA)
        buttonA.setDirection(Gpio.DIRECTION_IN)
        buttonA.setActiveType(Gpio.ACTIVE_LOW)
        buttonA.setEdgeTriggerType(Gpio.EDGE_BOTH)
        buttonA.registerGpioCallback {
            redLed.value = it.value
            true
        }
    }

    override fun onDestroy() {
        super.onDestroy()
        redLed.close()
        buttonA.close()
    }
}
```

It is exactly the same code we used for turning the red LED on when we pressed the button A. I purposely kept the same variable names and BoardDefaults; I also used the same pins on the diagram.

If you want to change the logic a bit (for example, turning the light off if we have not detected a presence for two minutes), you can still test that logic with an LED and a button, and then, once it works the way you want it, connect it to the proper hardware.

 Code the concept with buttons and LEDs, test the logic, then use the right components.

Note that if we want to have a system that turns on a fan for two minutes when a certain level of gas is reached, the code is still the same; it is just the components we connect that are different.

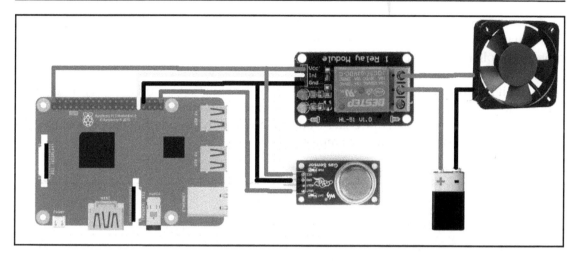

Using a smoke detector to trigger a fan

And, finally, let's look at other peripherals that use GPIO in less standard ways.

Other usages of GPIO

In many cases, circuits are required to use several digital signals as input or output. In these cases we need to know what they are expecting and handle it appropriately. This usually takes the form of a driver.

In this section we will be exploring briefly some of those cases, such as a DC motor controller that allows us to select the direction of two motors and a stepper motor controller.

We will also take a look at other non standard uses of GPIO that require special handling, such as an ultrasonic distance sensor and an LCD numeric display.

With all these, you will get a general idea of the type of peripherals that can be controlled using GPIO and the different ways they work.

DC motor controller (L298N)

A DC motor is a common component. If we just want to turn it on and off we can use a relay; but if we want to be able to change the direction we need something a bit more sophisticated: a motor controller.

Most motor controllers are designed to handle two DC motors, and the reason is that most cars will use two of them, one on each side, to be able to turn. We will be using the L298N motor controller in this example.

This circuit has four sets of connectors. A first set of six pins that include Vcc, Ground, and inputs 1 to 4 that are connected to the development kit, a connector for the main power source for the motors with an optional 5v output, and finally, two sets of connectors at each side of the board that connect to each motor.

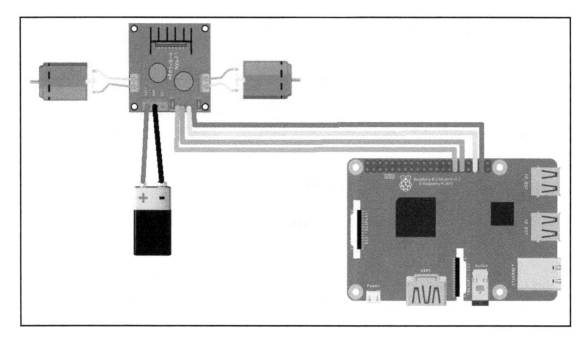

The four inputs are the ones that control the motors' direction. Each pair of pins controls one motor. If we set the first pin to high, the motor goes forward, and if we set the second pin, it goes backward.

A simple driver for one of the motors looks like this:

```
class Motor (gpioForwardPin: String, gpioBackwardPin: String) :
AutoCloseable {
    private val gpioForward: Gpio
    private val gpioBackward: Gpio

    init {
        val service = PeripheralManager.getInstance()
```

```
        gpioForward = service.openGpio(gpioForwardPin)
        gpioForward.setDirection(Gpio.DIRECTION_OUT_INITIALLY_LOW)
        gpioBackward = service.openGpio(gpioBackwardPin)
        gpioBackward.setDirection(Gpio.DIRECTION_OUT_INITIALLY_LOW)
    }

    override fun close() {
        gpioForward.close()
        gpioBackward.close()
    }

    fun forward() {
        gpioForward.value = true
        gpioBackward.value = false
    }

    fun backward() {
        gpioForward.value = false
        gpioBackward.value= true
    }

    fun stop() {
        gpioForward.value = false
        gpioBackward.value = false
    }
}
```

But you don't need to write the driver. Instead, we will include it as a dependency:

```
dependencies {
    [...]
    implementation 'com.plattysoft.things:l298n:+'
}
```

Opening a motor controller is quite simple; it just requires the name of the four GPIO pins:

```
motorController = L298N.open("BCM22", "BCM23", "BCM24", "BCM25")
```

When you combine the direction of the two motors, you can make the car go forward, go backward, turn left or right (moving one motor and keeping the other stopped), or spin (one motor forward, the other backward). The L298N class includes them as an enum named MotorMode with all the options.

You can simply call setMode on the motor controller:

```
motorController.setMode(MotorMode.TURN_LEFT)
```

And, with that, you have the basics of a robot car. You can make a self-driving car or just remote control it, be it with a keyboard or with a companion app, using either Bluetooth or simple server to provide an API. We will look at examples of communication with companion apps in the last chapter of the book.

Stepper motor (28BYJ-48) with controller (ULN2003)

Another interesting type of motor is the stepper motor. Together with DC motors and servos, they make the three types of motors that you see on most designs. We will talk about servos in the next chapter.

A stepper motor allows you to tell it how many steps you want it to take in a given direction. This allows for fine-tuned movement, and it can be used to move a latch on a door, activate a candy dispenser, roll curtains up and down, and so on:

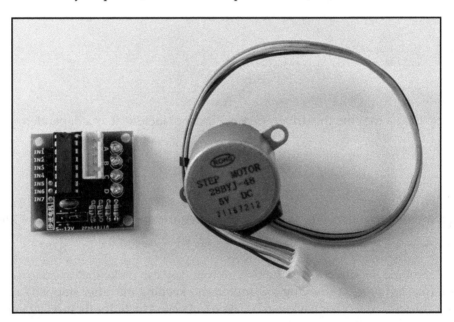

A stepper motor needs a controller circuit; a common one is ULN2003. The original version of this driver can be found at https://github.com/Polidea/Polithings but the maintainer is no longer active, so I have updated it to support Android Things 1.0 and uploaded it to PlattyThings. You can add this line to the dependencies of your project:

```
dependencies {
    [...]
    implementation 'com.plattysoft.things:uln2003:+'
}
```

As in the case of the DC motor controller, this component has four pins, but the usage is completely different; they are used to send signals to the motor to make it move. The way to make the motor take a step is to change the value of the input pins in a specific sequence. The driver is reasonably complex, so we won't be looking at it. However, using it is quite simple:

```
val stepper = ULN2003StepperMotor(in1Pin, in2Pin, in3Pin, in4Pin)

// Perform a rotation and add a rotation listener
stepper.rotate(degrees = 180.0,
        direction = Direction.CLOCKWISE,
        resolutionId = ULN2003Resolution.HALF.id,
        rpm = 2.5,
        rotationListener = object : RotationListener {
            override fun onStarted() {
                Log.i(TAG, "rotation started")
            }
            override fun onFinishedSuccessfully() {
                stepper.close()
            }
            override fun onFinishedWithError(degreesToRotate: Double,
rotatedDegrees: Double, exception: Exception) {
                Log.e(TAG, "error, degrees to rotate: {$degreesToRotate}
rotated degrees: {$rotatedDegrees}")
            }
        })
```

We just need to create an object of the type ULN2003StepperMotor, passing the name of the four GPIO pins and then call rotate with the amount of degrees, direction, resolution and **rpm (revolutions per minute)**. We also pass a listener to be notified of when the rotation is completed. In this simple example we use the listener to close the peripheral once the rotation is completed.

Typically stepper motors are not very fast because they go step by step counting the rotation until they are done. The model 28BYJ-48 can get up to 10 rpm, but not much faster than that.

Ultrasonic distance sensor (HC-SR04)

Another interesting component is a distance sensor. This particular component is interesting because of the way it uses GPIO.

The HC-SR04 has four pins: Vcc, Ground, echo, and trigger. Echo and trigger are GPIOs for input and output respectively. The sensor works by sending a signal via the trigger and then reading the response from the echo. The response has the information about the distance encoded in the duration of the pulse, it is proportional to how long the echo signal is on.

To use this sensor, we can include a community driver that is published to jcenter:

```
dependencies {
    [...]
    implementation 'com.leinardi.android.things:driver-hcsr04:+'
}
```

And then we can use it to read values in a fashion very similar to what we did for the temperature, in this case posting a runnable as fast as we can. Note that the readDistance method is synchronous; it sends the trigger and waits for the echo to return, so it can take up to 25 milliseconds to measure and return.

```kotlin
class DistanceSensorActivity : Activity() {
    companion object {
        [...]
    }

    lateinit var sensor: Hcsr04
    val handler = Handler()

    override fun onCreate(savedInstanceState: Bundle?) {
        super.onCreate(savedInstanceState)
        sensor = Hcsr04(triggerGpio, echoGpio)

        handler.post(object: Runnable {
            override fun run() {
                val distance = sensor.readDistance()
                Log.d("DistanceSensorActivity", "distance: $distance")
                handler.post(this)
            }
        })
    }

    override fun onDestroy() {
        super.onDestroy()
        sensor.close()
    }
}
```

Note that we have configured the pins as part of a companion object, similar to BoardDefaults, just as another example on how to include it in your classes. A companion object is a good match for small and simple tests.

The driver also supports sensor drivers. In that case, we need to add the MANAGE_SENSOR_DRIVERS permission to the manifest:

```xml
<uses-permission
android:name="com.google.android.things.permission.MANAGE_SENSOR_DRIVERS"
/>
```

And the code, as well, is very similar to the one we used to read the temperature. In this case, when the sensor registers, the parameter to the callback to `onDynamicSensorConnected` will have the type `TYPE_PROXIMITY`:

```
val sensorCallback = object : SensorManager.DynamicSensorCallback() {
    override fun onDynamicSensorConnected(sensor: Sensor?) {
        if (sensor?.type == Sensor.TYPE_PROXIMITY) {
            sensorManager.registerListener(sensorListener, sensor,
SensorManager.SENSOR_DELAY_FASTEST)
        }
    }
}
```

And inside the sensor listener, we will receive one value inside the values of the event object, which is the distance:

```
val sensorListener = object : SensorEventListener {
    override fun onSensorChanged(event: SensorEvent) {
        val distance = event.values[0]
        Log.i(ContentValues.TAG, "proximity changed: $distance")
    }

    override fun onAccuracyChanged(sensor: Sensor, accuracy: Int) {
        Log.i(ContentValues.TAG, "accuracy changed: $accuracy")
    }
|
```

The rest of the code includes the creation of a sensor driver, registering a callback with the sensor manager, and registering the proximity sensor. This is exactly the same as we did with the temperature and pressure sensors in the previous chapter:

```
sensorManager = getSystemService(Context.SENSOR_SERVICE) as SensorManager
sensorManager.registerDynamicSensorCallback(sensorCallback)
sensorDriver = Hcsr04SensorDriver(triggerGpio, echoGpio)
sensorDriver.registerProximitySensor()
```

And, of course, do not forget to unregister them when we are done:

```
sensorManager.unregisterDynamicSensorCallback(sensorCallback)
sensorDriver.unregisterProximitySensor()
```

Although, in theory, it should be as fast as continuous reading, this particular driver is a bit slow when used as a sensor driver. For a real-time critical system, such as a car avoiding obstacles, you are better using the first approach with the handler.

LCD display (TM1637)

One more interesting component that is actually included in the `contrib-drivers` package is the LCD display TM1637. This peripheral is very handy when we do not have something like the alphanumeric display of the Rainbow HAT at hand, and it is also a simple and cheap display you can add to your IoT projects.

To manage this peripheral, we just need to add one line to our dependencies on gradle:

```
dependencies {
    [...]
    implementation 'com.google.android.things.contrib:driver-tm1637:+'
}
```

The breakout circuit has four connectors: the typical Vcc and Ground, and then two labeled as signal and clock. Both extra pins are to be connected to GPIO, but they work in a similar fashion to I2C.

This circuit uses a one-way communication protocol that has a separate clock signal to synchronize the sender and the receiver, which usually allows for faster speeds as well. The value of the signal pin will change, and when the clock is active, the value can be read. Don't worry; all that is managed by the driver.

To see how to use this component, we are going to build a digital clock:

```
class LCDClockActivity : Activity() {

    companion object {
        [...]
    }

    val dateFormat = SimpleDateFormat("HHmm")
    val date = Date()
    lateinit var display: NumericDisplay

    override fun onCreate(savedInstanceState: Bundle?) {
        super.onCreate(savedInstanceState)
        display = NumericDisplay(dataGpioPinName, clockGpioPinName)
        display.setBrightness(NumericDisplay.MAX_BRIGHTNESS)

        Timer().schedule(timerTask {
            // Blink the colon
            display.setColonEnabled(!display.colonEnabled)
            // Update the values
            date.time = System.currentTimeMillis()
            display.display(dateFormat.format(date))
        }, 0, 1000)
    }

    override fun onDestroy() {
        super.onDestroy()
        display.close()
    }
}
```

As of now, this should be exactly what we expect: a constructor that receives the name of the two pins connected, some utility methods, and then closing the device inside onDestroy.

The sample does have a timer that runs every second. This time the exact execution of a timer is very important, since we will use it to make the : blink.

Inside the lambda of the timer task, we just change the enabled status of the : and then format the time using a simple date formatter to be shown on the display.

Summary

We have covered quite some ground in this chapter. First, we looked at how to address the pins from code in a way that it works on both developer boards. Then we learned about GPIO as output, how to access LEDs directly, and how to use relays. We also explored GPIO as input, taking off several abstraction layers from `Button` reaching direct access to GPIO, which we then used for other sensors. Finally, we learned about more general ways in which GPIO is used by other components, such as DC motor controllers, stepper motors, distance sensors, and numeric displays.

It is time to move on to another protocol that is a bit more complex but also allows us to do more: PWM.

PWM - Buzzers, Servos, and Analog Output

4

Pulse Width Modulation (PWM) is a one-wire, one-way communication protocol. Both Android Things developer kits have two PWM outputs -this is one case where Arduino has the upper hand- but we will see how to overcome that limitation in the next chapter. The Rainbow HAT uses one of them for the piezo buzzer and exposes the other via the pins on the HAT. We will take advantage of this setup by using some of the hardware of the Rainbow HAT for our examples.

 Arduino has a lot more PWM outputs than Android Things. We'll see how to work around that in the next chapter.

And as usual, since we will be using peripherals, we need to make sure that the USE_PERIPHERAL_IO permission is being declared in our manifest:

```
<uses-permission
android:name="com.google.android.things.permission.USE_PERIPHERAL_IO"/>
```

We will start by learning how it works and then explore its main basic uses: play sound with piezo buzzers, control servos, and use as analog output.

We will cover the following topics in this chapter :

- PWM overview
- Piezo buzzers
- Servos
- PWM as analog output

Technical requirements

You will be required to have Android Studio and Android Things installed on a developer kit. You also will require many hardware components to effectively perform the tasks given in this chapter. The components are very interesting to have, just to see them working, but the Rainbow HAT is particularly important. We go into details about the developer kits and how to pick the right one, as a part of Chapter 1, *Introducing Android Things*. Finally, to use the Git repository of this book, you need to install Git.

The code files of this chapter can be found on GitHub:
https://github.com/PacktPublishing/Android-Things-Quick-Start-Guide.

Check out the following video to see the code in action:

http://bit.ly/2oozzqr.

PWM overview

PWM is an output digital signal that is commonly used to apply a proportional control signal to an external device using a single digital output pin.

PWM is a square wave (no intermediate values, only 0 and 1) that oscillates according to a given frequency and duty cycle. It encodes the information in the length of the pulse, hence its name. A Pulse Width Modulation signal has two components:

- Frequency (expressed in Hz), which describes how often the output pulse repeats
- Duty cycle (expressed as a percentage), which describes the amount of time the pulse is on

For example, a PWM signal set to 50% duty is active for half of each cycle:

You can adjust the duty cycle to increase or decrease the amount of time the signal is on, which is reflected in the average value of it (the analog equivalent). The following diagram shows how the signal looks like with 0%, 25%, and 100% duty:

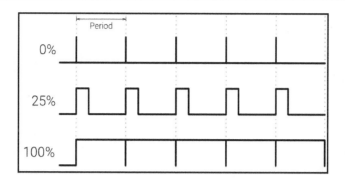

There are a few ways in which components read this information. We will be looking at them separately.

When we use the Rainbow HAT driver, the buzzer is connected to the second PWM pin, while the first one is designed to be used as a connection for a servo. Keep in mind that the naming of the pins depends on the developer kit: Raspberry Pi has them as PWM1 and PWM2, while iMX7D has them labeled PWM0 and PWM1. Both are configured on BoardDefaults of the driver, which we will be using in this chapter:

```
object BoardDefaults {
    private const val DEVICE_RPI3 = "rpi3"
    private const val DEVICE_IMX7D_PICO = "imx7d_pico"
    [...]
    val piezoPwm: String
        get() = when (Build.DEVICE) {
            DEVICE_RPI3 -> "PWM1"
            DEVICE_IMX7D_PICO -> "PWM2"
            else -> throw IllegalStateException("Unknown Build.DEVICE
${Build.DEVICE}")
        }
    val servoPwm: String
        get() = when (Build.DEVICE) {
            DEVICE_RPI3 -> "PWM0"
            DEVICE_IMX7D_PICO -> "PWM1"
            else -> throw IllegalStateException("Unknown Build.DEVICE
${Build.DEVICE}")
        }
    [...]
}
```

Let's start by accessing the piezo buzzer directly.

Piezo buzzers

Similar to what we did for GPIO, let's remove the Rainbow HAT driver from our dependencies and include only the drivers we require. For this first example, we just need to add the driver for the piezo buzzer, which is called `driver-pwmspeaker`:

```
dependencies {
    [...]
    implementation 'com.google.android.things.contrib:driver-pwmspeaker:+'
}
```

And then, we simply need to instantiate a `Speaker` object attached to the correct PWM pin:

```
class PianoActivity : Activity() {

    [...]

    override fun onCreate(savedInstanceState: Bundle?) {
        super.onCreate(savedInstanceState)
        buzzer = Speaker(BoardDefaults.piezoPwm)
        [...]
    }

    [...]
}
```

If we dig into the source code of the driver, we can see that it sets the duty cycle to 50% and then modifies the frequency when we invoke `play`. The buzzer uses the frequency from the PWM signal to vibrate and generate the sound.

If we want to bypass the driver and use PWM directly to control the buzzer, we can do it like this:

```
private lateinit var buzzer: Pwm

override fun onCreate(savedInstanceState: Bundle?) {
    super.onCreate(savedInstanceState)

    val pioService = PeripheralManager.getInstance()
    buzzer = pioService.openPwm(BoardDefaults.piezoPwm)
    buzzer.setPwmDutyCycle(50.0)

    [...]
}

override fun onDestroy() {
    super.onDestroy()
```

```
    buzzer.close()
    [...]
}
```

The initialization is now done using `PeripheralManager.openPwm()` and the only configuration is to set the duty cycle. Note that the `buzzer` variable type is now `Pwm` instead of `Speaker`.

Then, we can set the frequency and enable it to play as well as set `enabled` to `false` in order to make it stop.

 A PWM signal will stop completely when we disable it.

Regardless of the configuration of frequency and duty cycle, a PWM signal will stop completely when we disable it. We also need to enable it explicitly to make it send the pulse.

Note that we need to set the frequency before enabling the PWM pin; otherwise the operating system considers it as not properly initialized and throws an exception.

```
override fun onKeyDown(keyCode: Int, event: KeyEvent?): Boolean {
    val freqToPlay = frequencies.get(keyCode)
    if (freqToPlay != null) {
        buzzer.setPwmFrequencyHz(freqToPlay)
        buzzer.setEnabled(true)
        return true
    }
    else {
        return false
    }
}

override fun onKeyUp(keyCode: Int, event: KeyEvent?): Boolean {
    buzzer.setEnabled(false)
    return true
}
```

With these simple modifications, we can now control the piezo buzzer using the concepts of PWM. Let's move onto the most interesting components that you can control with this protocol: servo motors.

Servos

Besides relays, servo motors are one of the most interesting and versatile components that are frequently used on IoT projects. A servo motor can turn itself to a determined angle with high precision. They can be used to steer a robot car, orient a camera, as actuators on a robotic arm, and so on. They are -obviously- controlled using PWM.

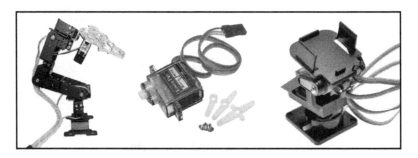

Robotic arms and camera brackets rely on servos

Servos use PWM in a particular way. They do not read the duty cycle in the way you would expect; they do check the length of the pulse. That length is not related to a percentage of the period, but an absolute number. A servo expects this value to be sent about every 20 ms, so we will work at 50 Hz. Repeating the signal is important to make the servo return to that position if it was moved -i.e. by brute force- but without an external interaction, the servo will simply stay in the last position.

A servo considers itself to have a neutral position when there is no left or right rotation, and that is indicated by a pulse of 1.5 ms in duration (note that it is nowhere close to a 50% duty cycle at 50 Hz, which is 10 ms).

 Servos use PWM in their own particular way.

The maximum and minimum pulse length varies slightly from each brand and even each particular servo, but typically they are around 1 ms for minimum and 2 ms for maximum:

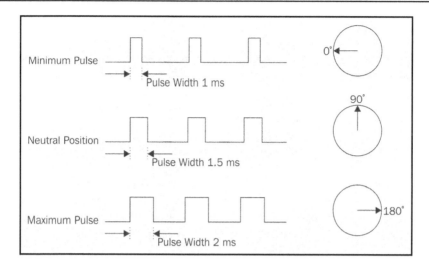

In my experience, most of the mini servos operate on a slightly wider pulse duration range. Do not worry; the servo driver allows us to configure and control its parameters easily, but it is important to understand what they mean. Let's build an example.

Using the servo driver

We will make an activity that allows us to select the target angle for a servo motor and apply it using components of the Rainbow HAT. To do so, we will use the buttons A and B to respectively decrease and increase the target angle, which we will display on the LCD. Finally, when we tap on button C, we will apply the target angle to the servo.

This sample project combines several components on the Rainbow HAT with a servo. It serves both as a reminder of how to use those components and as a showcase of how handy they can be when performing quick tests with other peripherals.

Since we are going to use a servo, buttons, and the alphanumeric display, we will need to add the following three dependencies to our `build.gradle`:

```
dependencies {
    [...]
    implementation 'com.google.android.things.contrib:driver-pwmservo:+'
    implementation 'com.google.android.things.contrib:driver-button:+'
    implementation 'com.google.android.things.contrib:driver-ht16k33:+'
}
```

The Rainbow HAT allows us to connect the servos really easily; it exposes the three pins (PWM, Vcc, and Ground) together and in the right order. Note that most servos come with a connector that just has these three pins. If you were to connect them to the developer kit directly, you would need extra wires.

When connecting the servo, use the colors for reference, especially the PWM, which is usually yellow.

When connecting the servo, you need to pay attention to the colors. Red and black should be used for Vcc and Ground respectively but not all manufacturers follow this standard (other similar colors, such as brown instead of black, may be used), but in all cases the PWM connection is yellowish and the Ground one is always darker:

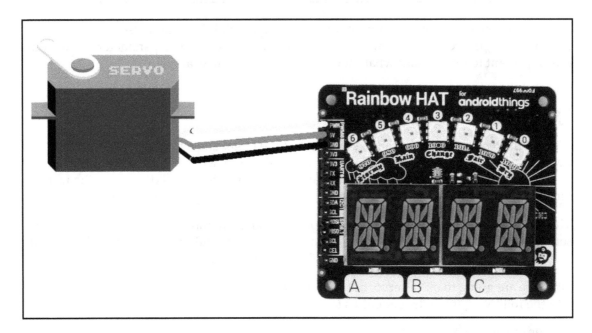

To initialize a servo we just have to instantiate an object of type `Servo` passing the right name for the PWM pin and then set the angle:

```
val servo = Servo(BoardDefaults.servoPwm)
servo.setEnabled(true)
servo.angle = targetAngle
```

In this case, we let the driver use the default values for the servo. Those values are 0 to 180 degrees for servo rotation and 1 to 2 milliseconds for pulse duration. Those defaults are fine for most servos, but always check the specs of your servo and modify them if required.

> The default values for the servo driver are 0 to 180 degrees for servo rotation and 1 to 2 milliseconds for pulse duration.

Also note that we need to enable the PWM output. It is the same behavior as with the buzzer: Android Things will not generate any output on that pin unless it is enabled.

> PWM only generates an output when it is enabled, so if you do not enable the pin, the servo won't move.

Let's explore the code of this example, starting with the member variables declaration:

```
class ServoActivity: Activity() {

    private lateinit var buttonA: ButtonInputDriver
    private lateinit var buttonB: ButtonInputDriver
    private lateinit var buttonC: ButtonInputDriver
    private lateinit var servo: Servo
    private lateinit var display: AlphanumericDisplay

    private var targetAngle = 0.0

    [...]
}
```

We will be using three button input drivers, the servo, and the alphanumeric display, so we just declare them all as `lateinit`, as usual. We also use a class variable named `targetAngle` to store the current angle selected.

Initialization is also quite straightforward:

```
override fun onCreate(savedInstanceState: Bundle?) {
    super.onCreate(savedInstanceState)

    servo = Servo(BoardDefaults.servoPwm)
    servo.setEnabled(true)

    display = AlphanumericDisplay(BoardDefaults.i2cBus)
    display.setEnabled(true)
```

```
display.setBrightness(AlphanumericDisplay.HT16K33_BRIGHTNESS_MAX)

buttonA = ButtonInputDriver(BoardDefaults.buttonA,
    PRESSED_WHEN_LOW, KEYCODE_A)
buttonB = ButtonInputDriver(BoardDefaults.buttonB,
    PRESSED_WHEN_LOW, KEYCODE_B)
buttonC = ButtonInputDriver(BoardDefaults.buttonC,
    PRESSED_WHEN_LOW, KEYCODE_C)

buttonA.register()
buttonB.register()
buttonC.register()
}
```

Inside `onCreate` we open the servo and enable it; the other components are initialized the same way as we did before. You may have noticed that the alphanumeric display uses the I2C bus. We will look into the details of that in the next chapter.

And we do the typical cleanup inside `onDestroy`:

```
override fun onDestroy() {
    super.onDestroy()

    servo.close()
    display.close()

    buttonA.unregister()
    buttonB.unregister()
    buttonC.unregister()
}
```

The reason why we are using `ButtonInputDriver` in this example is that it handles key repeats out of the box, and moving the value of the angle from 0 to 180 would be very tedious without that functionality.

We override both `onKeyDown` and `onKeyMultiple` using a common method to handle the key code and we pass the event up the chain if it wasn't handled:

```
override fun onKeyDown(keyCode: Int, event: KeyEvent?): Boolean {
    if (handleKeyCode(keyCode)) {
        return true
    }
    return super.onKeyDown(keyCode, event)
}

override fun onKeyMultiple(keyCode: Int, repeatCount: Int, event:
KeyEvent?): Boolean {
```

```
    if (handleKeyCode(keyCode)) {
        return true
    }
    return super.onKeyMultiple(keyCode, repeatCount, event)
}
```

Let's look at the code for handleKeyCode:

```
private fun handleKeyCode(keyCode: Int): Boolean {
    when (keyCode) {
        KEYCODE_A -> {
            decreaseAngle()
            return true
        }
        KEYCODE_B -> {
            increaseAngle()
            return true
        }
        KEYCODE_C -> {
            servo.angle = targetAngle
            return true
        }
        else -> {
            return false
        }
    }
}
```

As we mentioned before, to make the servo move to the specified position we just have to set its angle to the targetAngle and the driver will take care of it for us. Note that although Kotlin allows us to write the code as a variable assignment, it is calling a setter method behind the scenes, on which the PWM connection is configured.

 Kotlin allows us to set the angle as a variable assignment, but it is calling a setter method behind the scenes.

Finally, let's look inside the functions that increase and decrease the angle:

```
private fun decreaseAngle() {
    targetAngle--
    if (targetAngle < servo.minimumAngle) {
        targetAngle = servo.minimumAngle
    }
    display.display(targetAngle)
}
```

```
private fun increaseAngle() {
    targetAngle++
    if (targetAngle > servo.maximumAngle) {
        targetAngle = servo.maximumAngle
    }
    display.display(targetAngle)
}
```

Both are quite simple: first we increase or decrease the value until we reach the values the servo has configured as maximum or minimum angles, and then we display the value on the display.

With this example, we have a very simple and visual way to interact with a servo and check how it works. Let's take a look at how to modify the configuration.

Tweaking the servo configuration

Our example was using the minimum and maximum angles of the servo; we mentioned that the defaults are 0 to 180 degrees, but, can we change them? Obviously we can.

We can modify them with the setAngleRange method. In my experience, if we use the default pulse duration configuration most mini servos go just from 0 to 90 degrees. To reflect that we can use the following code:

```
servo.setAngleRange(0.0, 90.0)
```

There are many cases when we want to configure the angle range: smaller range -such as 0 to 90-, larger range -such as 0 to 360-, custom range -such as -90 to 90- for code clarity reasons, and so on.

The other configuration we can tweak is the range for the duration of the pulse. For that we have the setPulseDurationRange method. This range will create a map from the angles to the pulse duration. For the minimum angle the driver will send the shortest pulse, for the maximum angle it will send the longest pulse, and for the angles in between it will send a proportional pulse length.

 The default pulse duration range is 1 to 2 milliseconds, but most servos have a wider range.

The default values, as we mentioned earlier, are 1 to 2 milliseconds. That pulse duration is a safe bet, but you may want to play around with your servos. In the majority of the mini servos I have tested, the default values only cover up to 90 degrees, but a wider range allowed them to cover up to 180 degrees.

A less conservative pulse duration range that works in many servos is the following:

```
servo.setPulseDurationRange(0.65, 2.5)
```

But be careful when going over the default pulse duration range; each servo is slightly different.

> Play with the pulse duration range to find the best configuration for your servo.

With that, you should be able to properly configure pretty much any servo. Let's look at the last typical use of PWM: analog output.

PWM as analog output

In the introduction we talked about the average value of a PWM signal. That average can be considered an analog value for the signal as long as the frequency is high enough that the circuits do not notice it. Many circuits will have voltage regulators to compensate for the variations on the input and in that case it will be in fact an analog output. This analog value is directly proportional to the duty cycle.

Since we only have one PWM available we will use a single color LED to visually understand how analog outputs work. In the case of the LED, the intensity of the brightness will be proportional to the value of the duty cycle.

We will use a red LED for this example, connected as shown in the following diagram. We connect the LED to the PWM pin, add a 330 Ω resistor to take care of the extra voltage, and then we connect again to ground. A mini breadboard should be enough to do this circuit:

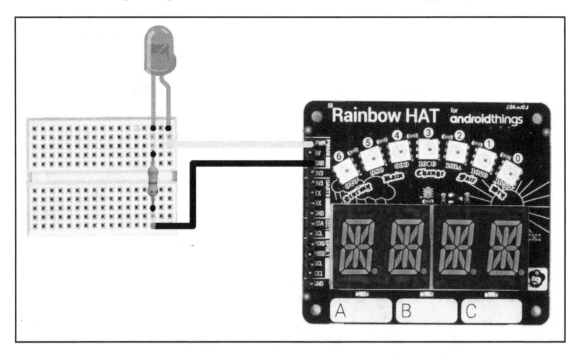

We could use this same circuit to blink the LED using GPIO, but, since it is connected to PWM, it will allow us to modify the voltage, which will have an impact on the intensity of the current passing through the resistor and hence on the intensity of the LED.

For this example, we will make the LED oscillate following a sinusoid signal. Let's look at the code:

```
class LedBrightnessActivity: Activity() {

    private val initialTime = System.currentTimeMillis()
    private lateinit var pwm: Pwm
    private val handler = Handler()

    private val ledRunnable = object : Runnable {
        override fun run() {
            val currentTime = System.currentTimeMillis()
            val elapsedSeconds = (currentTime - initialTime) / 1000.0
            val dutyCycle = Math.cos(elapsedSeconds) * 50.0 + 50
            pwm.setPwmDutyCycle(dutyCycle)
```

```
            handler.post(this)
        }
    }

    override fun onCreate(savedInstanceState: Bundle?) {
        super.onCreate(savedInstanceState)

        val pm = PeripheralManager.getInstance()
        pwm = pm.openPwm(BoardDefaults.servoPwm)
        pwm.setPwmFrequencyHz(50.0)
        pwm.setEnabled(true)

        handler.post(ledRunnable)
    }

    override fun onDestroy() {
        super.onDestroy()
        handler.removeCallbacks(ledRunnable)
        pwm.close()
    }
}
```

Inside onCreate we open the PWM using PeripheralManager, set the frequency to 50 Hz, and enable it. A frequency of 50 Hz is enough for the LED to assimilate the voltage as an average instead of a square wave. As usual, we have our cleanup code inside onDestroy.

Once again, remember that while the duty cycle has a default value, the frequency does not, and it must be set before calling setEnabled, otherwise the system will throw an exception.

In the same way as other examples in this book, we use a handler and a runnable to take some code and run it continuously without blocking the main thread.

The runnable itself calculates the value of the voltage based on a `cos` function with the center at 50%. The value of the duty cycle is shown in the following diagram. Note that the unit of x is radians:

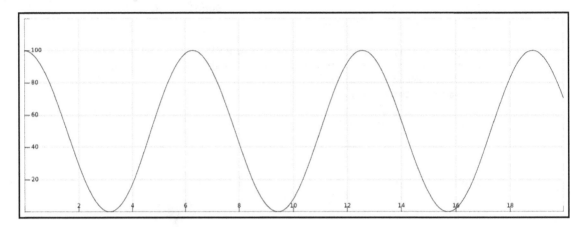

When we run this example, we can see how the intensity of the LED fluctuates following the timing of this chart, proving that PWM indeed works as an analog output.

Summary

In this chapter we have learned how PWM works and how it encodes information in both the frequency and the duty cycle. We have seen how the frequency is key to working with a passive buzzer and learned to manipulate it directly. We also learned how servos use PWM in a special way and how to modify the configuration to match the specs of a target servo. Finally, we have learned how to use PWM as analog output with a simple circuit.

So far we have used single direction protocols. Let's step up a level and start working with bidirectional communication with I2C, which also supports multiple devices on the same bus. This will be the focus of the next chapter.

5
I2C - Communicating with Other Circuits

The Inter-Integrated Circuit bus (also expressed as IIC, I²C, or I2C) is used to connect to simple peripheral devices with small data payloads, such as sensors, actuators, and simple displays. We will start with some general concepts about the protocol to then learn how to access I2C components from the Rainbow HAT directly. We will look at some extension components, displays, and other sensors that work with I2C. We will cover the following topics:

- Overview of I2C
- Revisiting Rainbow HAT components
- Extension components
- Small displays
- Other I2C sensors

As with PWM, the Rainbow HAT exposes the pins, so we can use the hardware on the HAT to interact with the new extra components we plug into the I2C bus.

Technical requirements

You will be required to have Android Studio and Android Things installed on a developer kit. You also will require many hardware components to effectively perform the tasks given in this chapter. The components are very interesting to have, just to see them working, but the Rainbow HAT is particularly important. We go into details about the developer kits and how to pick the right one, as a part of `Chapter 1`, *Introducing Android Things*. Finally, to use the Git repository of this book, you need to install Git.

The code files of this chapter can be found on GitHub:
`https://github.com/PacktPublishing/Android-Things-Quick-Start-Guide`.

Check out the following video to see the code in action:

`http://bit.ly/2N75Sbg.`

Overview of I2C

I2C is a synchronous serial interface, which means it uses a shared clock signal to synchronize data transfer between devices. The device that controls the clock signal is known as the master; in our case, the developer kit. All other connected peripherals are known as slaves. Each device is connected to the same set of data signals to form a bus.

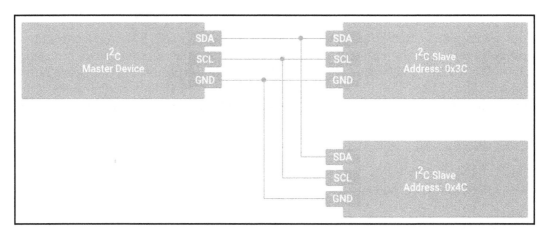

I2C structure diagram from developers.google.com

Being a serial bus, the components need to wait for their turn to talk. That is also done by the master, which issues commands to the bus and some of those commands may require a response (for example, requests to read values). This is known as half-duplex communication.

I2C devices connect using a three-wire interface consisting of the following pins:

- **Shared clock signal** (**SCL**): The clock used to synchronize the data transfer
- **Shared data line** (**SDA**): The wire used to actually communicate the data
- **Common ground reference** (**GND**): This pin is missing on the Rainbow HAT section of I2C, basically because you can use any ground pin that is already connected -for example, to power the components- so that ground connection is shared

Note that I2C can have multiple slaves, so they need a way to know which one is the master talking to. That is done using the addressing structure that is part of the I2C protocol.

Addressing circuits

Each device supporting I2C is programmed with a unique address and only responds to transmissions the master sends to that address. Every slave must have an address and it will be used on every transmission, even if the bus contains only a single slave.

We already saw these addresses when we looked at the back of the Rainbow HAT in Chapter 2, *The Rainbow HAT*. The I2C devices had them specified there:

- BMP280: I2C1, 0x77
- HT16K33L I2C1, 0x70

Those are precisely the addresses of the slaves. It is common for I2C-enabled components to have a base address and allow the last bits of it to be configured via hardware (address pins) so more than one device of the same type can be connected to the same I2C bus.

BYTE	BIT							
	7(MSB)	6	5	4	3	2	1	0 (LSB)
I²C slave address	L	H	L	L	A2	A1	A0	R/$\overline{W}$

Screenshot from pcf8575 spec sheet where we see the address composition

 The I2C bus has the same value on both Android Things developer kits: I2C1.

The Raspberry Pi and the iMX7D have the same value for the I2C bus. We can still have it as part of BoardDefaults, but it is reasonably safe to have it as a constant.

```
val i2cBus: String
    get() = when (Build.DEVICE) {
        DEVICE_RPI3 -> "I2C1"
        DEVICE_IMX7D_PICO -> "I2C1"
        else -> throw IllegalStateException("Unknown Build.DEVICE
${Build.DEVICE}")
    }
```

In any case, you can always find the default I2C bus of a board via the `PeripheralManager` in the following way:

```
fun getI2CBus(): String {
    val peripheralManager = PeripheralManager.getInstance()
    val deviceList = peripheralManager.i2cBusList
    if (!deviceList.isEmpty()) {
        return deviceList[0]
    }
    return "I2C1"
}
```

In general, drivers abstract us from the use of the peripheral manager and we just need to build an instance of the object or call `open`, depending of the flavor. Most of the existing drivers have multiple variants of the open method/constructors; one of them usually that has no parameters and takes the default I2C bus and the default base address of the component, to make connecting to those devices easy.

However, if at any time you need to find out the address of an I2C peripheral manually, there is a simple method to do so:

```
fun scanAvailableAddresses(i2cName: String): List<Int> {
    val pm = PeripheralManager.getInstance()
    val availableAddresses = mutableListOf<Int>()
    for (address in 0..127) {
        val device = pm.openI2cDevice(i2cName, address)
        try {
            device.write(ByteArray(1), 1)
            availableAddresses.add(address)
        } catch (e: IOException) {
            // Not available, not adding it
        } finally {
            device.close()
        }
    }
    return availableAddresses
}
```

This function will loop for each address and try to write a byte to it. If it succeeds, then it means a device is connected. The function will return a list of detected device addresses.

And as always, remember that you need to have the `USE_PERIPHERAL_IO` permission declared on your manifest:

```
<uses-permission
android:name="com.google.android.things.permission.USE_PERIPHERAL_IO"/>
```

Revisiting Rainbow HAT components

Following the same approach as for GPIO and PWM, we are going to start by removing the Rainbow HAT meta driver from our dependencies and using the specific drivers of the components directly. This time the components are the alphanumeric display and the temperature and pressure sensor.

Alphanumeric display (Ht16k33)

The driver for the alphanumeric display is called `driver-ht16k33`, so we just need to add it to our `build.gradle`:

```
dependencies {
    [...]
    implementation 'com.google.android.things.contrib:driver-ht16k33:+'
}
```

The only difference with the code we used before is that we now have to construct an object of type `AlphanumericDisplay` passing the I2C bus name, which we can initialize using one of the patterns described earlier; either `BoardDefaults` or `getI2CBus()`.

```
val display = AlphanumericDisplay(i2cBusName)
display.setEnabled(true)
[...]
```

And that's it; the rest of the interaction with the alphanumeric display is exactly the same as we did before.

Temperature and pressure sensor (Bmx280)

In a similar way, for the temperature and pressure sensor we just need to include the sensor driver into the `build.gradle`:

```
dependencies {
    [...]
    implementation 'com.google.android.things.contrib:driver-bmx280:+'
}
```

And then build an object of type Bmx280, passing the bus name:

```
val bmx280 = Bmx280(i2cBusName)
bmx280.setTemperatureOversampling(Bmx280.OVERSAMPLING_1X)
bmx280.setMode(Bmx280.MODE_NORMAL)
[...]
```

Also very simple. When using custom drivers most of the complexity is understanding the driver itself.

Note that both of the components are connected to the same bus and we can interact with both within the same activity, as we did when we displayed the temperature on the display. That is why the I2C protocol is very interesting. Instead of holding the pins for a particular circuit, it allows us to connect multiple devices to the same shared bus.

Let's look at some other drivers for components that we can use to extend the functionality of the developer kits.

Extension components

In this section we will talk about some components that extend the functionalities of our developer kits, allowing us to do more and overcoming some of the deficits with Arduino, such as the lack of analog inputs and the very limited amount of PWM pins.

Analog to digital converter (ADC) – PCF8591

There are many sensors that just provide an analog value: moisture, smoke, raindrop, light, temperature, and so on, and as we mentioned several times, the Android Things developer kits do not have any analog inputs. To be able to use these sensors as input, we need a special component called an Analog to Digital Converter (ADC).

An ADC reads an analog value and converts it to a digital value. The resolution depends on the chip; most chips offer 8 or 10 bits.

There is a driver for an ADC in the contrib-drivers library, but the component itself has been retired, so we will be using a different one: PCF8591.

 The ADC included in the contrib-drivers set is no longer available

This chip has four analog input channels with eight bits of resolution, that are values between 0 and 255. Many breakout circuits based on this chip include a potentiometer, a thermistor, and a light resistor on the board; which can be connected to the analog inputs using jumpers. I find that quite handy to be used as example without needing any extra wiring.

For example, in the breakout circuit, the potentiometer is **INPUT0**, the light resistor is **INPUT1**, and the thermistor is **INPUT2**, and they can easily be connected to **AIN2** (AIN stands for Analog Input,) respectively:

A breakout circuit of PCF8591 with a potentiometer, a light variable resistor, and a thermistor included on the board

To use this ADC, we just need to add the driver to our list of dependencies:

```
dependencies {
    [...]
    implementation 'com.plattysoft.things:pcf8591:+'
}
```

Then we can just invoke `Pcf8591.open()` to get an instance of the driver mapped to the default I2C bus and address:

```
private val pcf8591 = Pcf8591.open()
```

Once the `Pcf8591` object is created, you can read the value from any analog input just by calling `readValue` and passing the channel name, or `readAllValues` to get them all in an array. Values are in the range from 0 to 255 (eight bits of resolution).

```
// Reading the channel from AIN2
val value = pcf8591.readValue(2)
// Reading the value from all channels
val allValues = pcf8591.readAllValues()
```

So, that is how we read analog values, but what kind of sensors can we read from?

Analog sensors

As we have mentioned before, there are many analog sensors that we can use. We talked about raindrop and smoke sensors in the GPIO chapter because they have a digital output that works as a trigger, but those sensors also provide an analog output. There are many other components we can also use, such as light dependent resistors, moisture sensors, or just potentiometers.

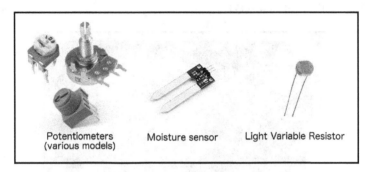

Potentiometers (various models) Moisture sensor Light Variable Resistor

In most cases, an analog sensor works just as a variable resistor that varies depending on the characteristic you want to measure, be it temperature, light, CO_2 particles in the air, water, and so on.

Also in most cases, the variability of the resistance is not linear, but has the shape of a $1/x$ function. That is caused by the fact that you can't get 0 or infinite resistance. You can see a couple of examples in the following diagram:

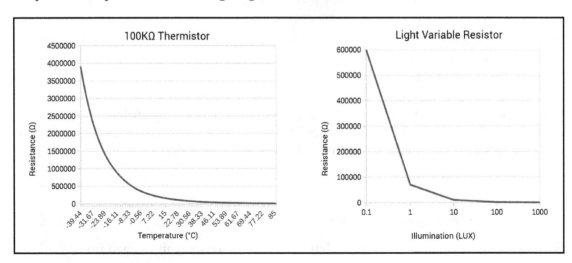

The most common way to use analog sensors is to measure a few checkpoints beforehand and use them as triggers. Light variable resistors, for example, are commonly used to decide to turn on the lights when it is dark enough, a moisture sensor could open a tap to water a plant when it is getting too dry, and so on.

 When working with analog sensors, you can use a potentiometer to test the logic and then replace it with the real sensor.

In the same way as you can replace any GPIO sensor with a button and any single GPIO output with a LED to test the logic, you can replace any analog sensor with a potentiometer, which is handier and also safer (imagine that you are testing a CO2 alarm).

PWM expander – PCA9685

The other big limitation of Android Things developer kits is that they only have two PWM outputs. For a robotic arm or just an RGB LED we need at least three.

The PCA9685 chip is a PWM expander. The original version of the driver is a bit too rough and is not a good example of a driver, so I have made a few improvements to the interface. For example, the original version does not use the default I2C address of the component; neither does it provide an interface consistent with the PWM one for setting the duty cycle.

To add the driver to your project, you need to add this line to the dependencies in the `build.gradle` of the module:

```
dependencies {
    [...]
    implementation 'com.plattysoft.things:pca9685:+'
}
```

We will build an example that uses this component together with the ADC we just saw. We are going to set the value of a single RGB LED using PWM as analog output for each of the colors, and the value for each color will come from a potentiometer.

This example makes use of the two components we need to overcome: the lack of analog inputs and outputs on Android Things. Funny enough, this particular example does not need any extra hardware when done on Arduino.

This is an almost trivial example on Arduino, but it requires two extra components on Android Things.

Although we are going to interact with a RGB LED, we could just as well use the potentiometers to control a robot hand using servos, and the code would be essentially the same. We'll use an RGB LED because it is way cheaper than a robotic arm, but I acknowledge that it is also significantly less impressive.

Let's take a look at the code:

```
class RgbLedActivity: Activity() {

    private lateinit var pwmExpander: PCA9685
    private lateinit var adc: Pcf8591
    private val handler = Handler()

    override fun onCreate(savedInstanceState: Bundle?) {
        super.onCreate(savedInstanceState)
        adc = Pcf8591.open()
        pwmExpander = PCA9685()

        handler.post(object : Runnable {
            override fun run() {
                val values = adc.readAllValues()
                for (i in 0..2) {
                    pwmExpander.setPwmDutyCycle(i, (values[i] /
256.0).toInt())
                }
                handler.postDelayed(this, 200)
            }
        })
    }

    override fun onDestroy() {
        super.onDestroy()
        adc.close()
        pwmExpander.close()
    }
}
```

It is reasonably simple: we open both components and then we set a handler to read as fast as we can from the ADC. On every read we take the values we have read -which are in the range from 0 to 255- and convert them to a percentage, which we then set to the specific PWM channel.

Note that each driver uses a slightly different flavor; we use `open` to obtain the `Pcf8591` object and a constructor to obtain the `PCA9685` one.

On the PWM expander side, note that we are not setting the frequency, the driver is setting it to 50Hz by default. A small limitation of this controller is that all PWM pins work at the same frequency, which is not a big deal, but it is important to be aware of.

 The frequency setting for the PCA9685 chip is shared among all the PWM pins.

Now for the interesting part; let's look at the wiring diagram:

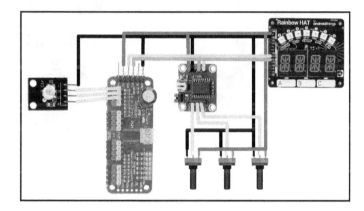

It is worth highlighting again that both circuits are connected to the same wires for I2C; the bus supports multiple slaves and the components are addressed by software. Also note that the ground and Vcc connections are used by a lot of elements. It does look quite neat on a diagram, but this simple circuit can become quite messy when wired physically.

And that's it. This is the only complex wiring you will come across in this book.

A final note about this component is that when you want to use it as a servo controller, there is a special class -`ServoUnderPca9685`- that allows us to control a servo with the same interface than the one in `contrib-drivers`.

GPIO expander – PCF8575

It may happen that we do not have enough GPIO pins, or they are not exposed, like when using the Rainbow HAT. For those cases, we can use a GPIO expander. The PCF8575 chip provides 16 GPIO ports (P00-P07 and P10-P17) that we can access via the I2C bus.

As usual, you need to add the driver to the module specific `build.gradle` dependencies:

```
dependencies {
    [...]
    implementation 'com.plattysoft.things:pcf8575:+@aar'
}
```

And then you can open the driver using the default bus and base address as well as specifying them if you need to:

```
private val gpioBoard = Pcf8575.open()
```

This chip has three pins named A0, A1, and A2, which can be used to configure the lower part of the I2C address, so you can connect more than one of these components on the same bus. In the breakout circuit, all of them are connected to Ground, but you can change that if you need to.

Once the `Pcf8575` object is created, you can open GPIO pins by name and use them like any other GPIO. The `openGpio` method returns objects that implement the `Gpio` interface.

```
// Opening a GPIO for an LED
val led = gpioBoard.openGpio("P00")
led.setDirection(Gpio.DIRECTION_OUT_INITIALLY_LOW)
led.setValue(true)
```

If you want to receive GPIO callbacks, you need to configure the interrupt pin. That pin is a GPIO that triggers when any of the GPIO pins configured as input changes. Since the interrupt pin is shared, it triggers for each change on any pin, so it is very noisy. You may want to implement some debouncing and filtering (or open it using the `button` driver) if you plan to use this chip for reading inputs.

 If you don't configure the interrupt pin, the callbacks will not be triggered.

You can configure the interrupt pin like this:

```
// Opening a GPIO for reading
val button = gpioBoard.openGpio("P00")
button.setDirection(Gpio.DIRECTION_IN)
val value = button.getValue()

// First configure the interrupt pin
gpioBoard.setInterrupt(interruptPin)

// Now we can register a callback
button.registerGpioCallback(GpioCallback() {
    Log.d("PCF8575","Read value (INT): ${it.getValue()}")
    true
})
```

Small displays

Another typical component for IoT projects are simple displays. We have already used two different LED segment-based displays, but there are other models that offer more functionality and versatility while still being cheap, such as LCD displays and OLED screens.

Both the LCD screen and the OLED screen drivers work by giving us access to the pixels on the screen, including some utility methods to draw text or even bitmaps depending on the case. The main difference is that on the LCD screen, the pixels are grouped into separated characters (5 by 8 pixels each), arranged in rows and columns (typically, two rows of 16 characters), while the OLED is a continuous screen, typically monochrome.

2x16 characters LED Screen   128x64 pixels OLED Screen

To use an LCD screen, we need to add the following driver by Gautier Mechling:

```
dependencies {
    [...]
    implementation 'com.nilhcem.androidthings:driver-lcd-pcf8574:+'
}
```

And then we can instantiate an object of type `LcdPcf8574` and use it to display text. This is a simple use of the driver:

```
val lcd = LcdPcf8574(i2cBusName, I2C_ADDRESS)
lcd.begin(16, 2)
lcd.setBacklight(true)

lcd.clear()
lcd.print("Hello,")
lcd.setCursor(0, 1)
lcd.print("Android Things!")

lcd.close()
```

It has functions to clear the screen and to print letters on it as well as to set the cursor to a specific position. It is a simple and effective way to display text.

Although we are using the LCD display with a I2C controller, it can also be controlled directly with GPIO inputs; but I don't recommend it, as there is too much wiring. The I2C model comes with a small converter chip, very similar to the GPIO expander we saw earlier. There are several components that fit this role (PCF8574, LCM1602, FC-113, and so on) and each model has its own I2C address (it is manufacturer specific). You can check the specs or simply use the scanAvailableAddresses utility method we saw in the first section to find the address.

On the other hand, the OLED screen can be wired to I2C as well as SPI (we will see that on the next chapter). Pay attention to the description of the component to make sure you are getting the version you want.

 Many display-type components can work on I2C and SPI, but when included on a breakout circuit, they are wired for either I2C or SPI.

The driver for the OLED screen is called SSD1306, and it is part of the contrib-drivers repository managed by Google:

```
dependencies {
    [...]
    implementation 'com.google.android.things.contrib:driver-ssd1306:+'
}
```

This driver allows us to set each pixel to on or off using setPixel and then call show to make it render the screen. Let's look at a simple example:

```
val display = Ssd1306(i2cBusName)
for (i in 0 until display.lcdWidth) {
    for (j in 0 until display.lcdHeight) {
        display.setPixel(i, j, i%display.lcdHeight > j)
    }
}
// render the pixel data
display.show()
// Cleanup
display.close()
```

The Ssd1306 class has multiple constructors, including some that have the width and the height of the display as parameters. By default it configures a display of 128x64 pixels. It also has methods to clear the screen, set it on or off, configure the contrast, and even trigger scrolling.

All those simple screens are not very expensive and can be handy in many IoT projects.

To complete our review of I2C peripherals, let's talk about other sensors.

Other I2C sensors

There are many variations of magnetometers, accelerometers, gyroscopes, and so on. In some cases, the component has a single function, such as the module HMC5883L, which is just a magnetometer, or ADXL345 and MMA7660FC (part of `contrib-drivers`), which are accelerometers. In other cases, they are more complete, even proper 9 DoF (Degrees of Freedom) sensors like the ones on the phones, which are also known as IMUs (Inertial Measurement Unit).

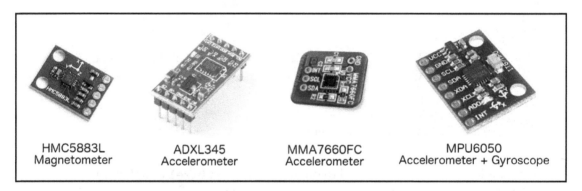

| HMC5883L | ADXL345 | MMA7660FC | MPU6050 |
| Magnetometer | Accelerometer | Accelerometer | Accelerometer + Gyroscope |

Check whether a driver exists before deciding on a component, or be ready to write your own.

In most cases there are drivers already available, even if most of them are just ports from Arduino. The structure of the driver varies a bit but it is usually along the lines of what we saw in this chapter. In the case that there is no driver, you can always read the spec sheet of the component and write it yourself. It is not as complicated as it sounds, but nevertheless, it is a good idea to check for drivers before buying any component.

In general, I2C drivers provide easy-to-read variables and method names and simplify the commands you can send to the peripheral, providing an abstraction over the formatting of the communication protocol, but it is common that they just mirror whatever protocol the component has.

As a reference, we are going to look at the MPU6050 component, which is an accelerometer plus gyroscope (6 DoF).

Accelerometer/gyroscope – MPU6050

This peripheral is a good example of a chip that gives us good resolution for orientation and acceleration at a fair price. The breakout circuit has many pins in addition to the ones we are used to. It has extra pins for SDA and SCL that can be used to interface other I2C modules with the MPU6050, a chip selection one (AD0) which can be used to modify the I2C address, and an interrupt one which can be used to read from the component on demand, similarly to what we saw with the GPIO expander.

For this example we will only use the basic four: Vcc, Ground, SCL, and SDA.

To get the driver, we need to add it to the module dependencies:

```
dependencies {
    [...]
    implementation 'com.plattysoft.things:mpu6050:+'
}
```

The driver itself is quite straightforward. I modified the original version slightly to adhere to the standard methods for opening the component we have been using. A simple usage is like this:

```
val gyroscope = Mpu6050.open()
Log.d(TAG, "Accel: x:${gyroscope.accelX} y:${gyroscope.accelY}
z:${gyroscope.accelZ}")
Log.d(TAG, "Gyro: x:${gyroscope.gyroX} y:${gyroscope.gyroY}
z:${gyroscope.gyroZ}")
gyroscope.close()
```

The driver has the values for acceleration as `accelX`, `Y`, and `Z` and, similarly, gyroscope orientation is presented as `gyroX`, `Y`, and `Z`.

 Kotlin allows us to use getters as just variable reads. Keep in mind that in most drivers those getters are synchronous blocking operations.

Note that, one more time, these are getters that issue a request to the component for reading and then receive a response. They are synchronous and can take a bit of time to execute, but Kotlin allows us to express them just as variables.

Summary

In this chapter we have covered the I2C protocol, which allows us to communicate with more complex peripherals. We have looked at the pin structure and how it works internally to address multiple components into the same bus.

Then we looked at the components of the Rainbow HAT that use I2C (alphanumeric display and temperature sensor) to realize that the meta driver layer is really thin. It is just hiding one parameter and the constructor. The complexity of handling these type of circuits is all about learning how each driver works.

We also looked at a few expansion boards: Analog to Digital Converter (ADC), PWM, and GPIO expanders. With all those, we can overcome the lack of analog inputs and outputs of the developer kits of Android Things.

Finally, we checked other type of sensors, including magnetometers, accelerometers, and gyroscopes, which are common in phones nowadays.

It is now time to explore the last of the basic protocols, designed for faster communication than I2C: SPI - Serial Parallel Interface.

SPI - Faster Bidirectional Communication

6

SPI is the last of the protocols to handle peripherals that we will be exploring in this book. It is similar to I2C, but also has a few differences, so we will start with an overview of the protocol. As usual, we will look at the component of the Rainbow HAT that is handled using the protocol -in this case, the RGB LED strip-. Finally, we will look at a couple of examples where SPI is commonly used: displays.

The topics covered in this chapter are:

- Overview of SPI
- LED strip
- Usage on displays

Let's start by learning about the protocol itself.

Technical requirements

You will be required to have Android Studio and Android Things installed on a developer kit. You also will require many hardware components to effectively perform the tasks given in this chapter. The components are very interesting to have, just to see them working, but the Rainbow HAT is particularly important. We go into details about the developer kits and how to pick the right one, as a part of Chapter 1, *Introducing Android Things*. Finally, to use the Git repository of this book, you need to install Git.

The code files of this chapter can be found on GitHub:
`https://github.com/PacktPublishing/Android-Things-Quick-Start-Guide`.

Check out the following video to see the code in action:

`http://bit.ly/2MGJ2bi`.

Overview of SPI

Serial Peripheral Interface (**SPI**) is typically used on external non-volatile memory and graphical displays, where faster data transfer rates and high bandwidth are required. Many sensor devices support SPI in addition to I2C.

SPI shares some characteristics with I2C. It is a synchronous serial interface, which means it relies on a shared clock signal to synchronize data transfer between devices. Because of the shared clock signal, SPI also has a master-slave architecture, where the master controls the triggering of the clock signal and all other connected peripherals are slaves.

> SPI is faster than I2C. The clock signal for data transfer is typically in the range of 16 MHz to 25 MHz.

Another similarity with I2C is that it supports multiple slaves.

But there are also some differences. While I2C is semi-duplex, SPI supports full-duplex data transfer, meaning the master and slave can simultaneously exchange information. This is achieved by having a wire to transmit from the master to the slaves and another one from the slaves to the master, which makes SPI a minimum 4-wire interface:

- **Master Out Slave In** (**MOSI**)
- **Master In Slave Out** (**MISO**)

- **Shared Clock Signal (CLK or SCL)**
- **Common Ground Reference (GND)**

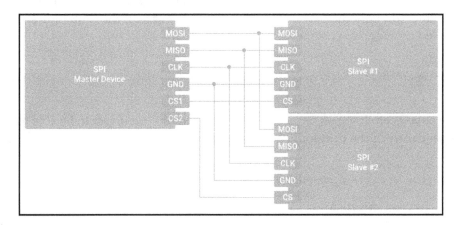

SPI structure diagram from developers.google.com

The other key difference is that the slaves are addressed via hardware instead of software. Each SPI slave has a chip select pin to allow the master to address that particular device. Addressing is necessary for data transfer, since all the slaves share the same bus. Chip select is not mandatory when the bus has a single slave.

Sometimes, SPI breakout circuits do not expose the CS. In that case, that peripheral needs to be the only slave on the bus.

And, as for any example with peripherals, when using SPI we need to add the required permission to the manifest:

```
<uses-permission
android:name="com.google.android.things.permission.USE_PERIPHERAL_IO" />
```

LED strip

As we have done in previous chapters, let's remove the dependency of the Rainbow HAT meta driver and add only the driver that we need. In the case of the RGB LED strip, that driver is called `driver-apa102`:

```
dependencies {
    implementation 'com.google.android.things.contrib:driver-apa102:+'
}
```

And, as we have been doing so far, let's use a `BoardDefaults` object to be able to run our code independently of the developer kit pinout naming:

```
val spiBus: String
    get() = when (Build.DEVICE) {
        DEVICE_RPI3 -> "SPI0.0"
        DEVICE_IMX7D_PICO -> "SPI3.1"
        else -> throw IllegalStateException("Unknown Build.DEVICE
${Build.DEVICE}")
    }
```

There are a couple of important things to note here. First is that the exposed SPI bus on the Raspberry Pi is called 0, while the iMX7D one is 3. It is also important to note that the order of the chip select pins on both boards is reversed: SS1 and SS0 are swapped. That is why the iMX7D has the bus `SPI3.1`, while the Raspberry Pi has `SPI0.0` when using the Rainbow HAT pins.

 The address pins on the master can have different labels: `SS0/1 -Slave Select-`, `CS0/1 -Chip Select-` or `CE0/1 -Chip Enable.`

An important part of naming the bus on SPI is that since the slaves are addressed by hardware, and the system provides us with a specific bus ID to talk to each slave. So, when we use `SPI0.0`, we access the slave associated with CS0 on bus SPI0 and, when we use `SPI0.1`, we access the slave associated with CS1 (SPI3.0 and SPI3.1, in the case of the iMX7D). We do not need to do anything else; all the addressing is done at the operating system level.

 The operating system provides us with a specific bus ID to talk to each slave: `SPI0.0` and `SPI0.1`, in the case of a Raspberry Pi.

Finally, as happened with I2C, the only change we need to do to the code is to replace the utility method of `RainbowHat` with the constructor passing the correct bus name:

```
mApa102 = new Apa102(spiBusName, Apa102.Mode.BGR);
```

All the complexity of handling SPI devices comes from the communication with the peripheral itself, which is reflected in the driver.

This one was easy. Let's look at some other drivers.

Usage on displays

We used an LCD display and an OLED screen in the I2C chapter. This time we will look at an LED matrix and the same OLED screen again, but wired to SPI so that you can see the differences:

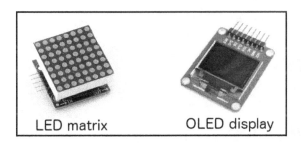

Note that the Rainbow HAT does have SPI pins exposed. Be aware that early versions of the Rainbow HAT have these pins wrongly labeled (the wiring is consistent, just the label is incorrect). Double check the correct labels or just connect the peripherals directly to your developer board:

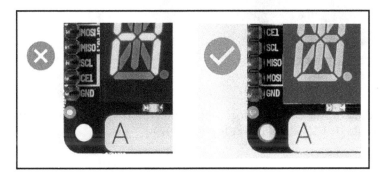

Incorrect and correct labeling of the SPI pins on the Rainbow HAT

Double-check the SPI pin labels on the Rainbow HAT, as early versions have the incorrect label.

LED matrix

This peripheral is an 8x8 LED matrix. It is great for retro-style UI, but obviously it is also very limited.

The names on the pins are slightly different than the ones on the board, so let's look at the connection diagram:

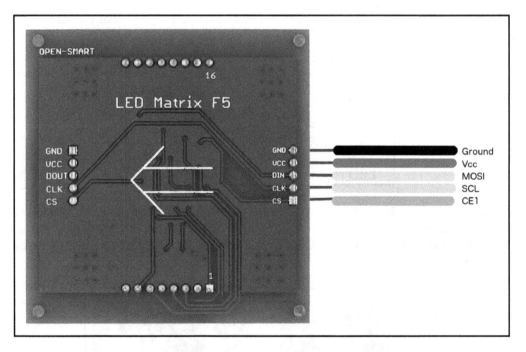

Ground and **Vcc** are the same as always.

DIN stands for data input, and therefore it is connected to the pin that transmits data from the master to the slave: **MOSI** (Master Out, Slave In).

SCL stands for shared clock, and **CLK** stands just for clock. The different naming can be confusing, but both are commonly used, so just be aware of them being the same.

Finally, we have **CS** (chip select). This is the pin that is used for addressing the peripheral, so, depending on where you connect it, your SPI bus will have one name or another. For example, if you connect CS to CE1 on a Raspberry Pi, then the bus will be SPI0.1, and if you connect CS to CE0, the bus will be SPI0.0.

One thing that stands out with from breakout circuit is that it has output pins on the left-hand side in addition to the input pins on the right. That is to allow chaining; you can connect another matrix to those pins and they will all work together. And not just two; you can chain any number of them (limited by the controller's addressing capability, but quite large anyway). If you want to have rows, you just need to connect the last one from one row to the first one on the next row.

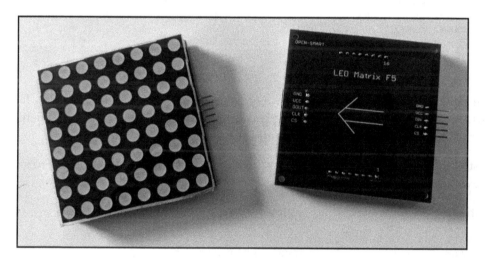

Now that we have the hardware side sorted, let's go into the software. First we need to add a driver to the dependencies on the `build.gradle` of our module:

```
dependencies {
    [...]
    implementation 'com.nilhcem.androidthings:driver-max72xx:+'
}
```

We mentioned that this is ideal for retro style UI, so we will be doing an animated sprite from `Space Invaders`. The code looks like this:

```
private lateinit var ledControl: LedControl
private val timer = Timer()

override fun onCreate(savedInstanceState: Bundle?) {
    super.onCreate(savedInstanceState)
    ledControl = LedControl(SPI_BUS, 1)
    initLedMatrix()
    timer.schedule(timerTask {
        swapSprite()
    }, 1000, 1000)
}

override fun onDestroy() {
    super.onDestroy()
    timer.cancel()
    timer.purge()
    ledControl.close()
}
```

Typical initialization is located inside `onCreate` and cleanup inside `onDestroy`. The constructor of `LedControl` is interesting because it has a second parameter, which is the number of chained matrices. For this example, we only use one (8x8).

The rest of the initialization is inside `initLedMatrix` and then we use a timer to swap the sprite every second.

The initialization of the matrix is as follows:

```
private fun initLedMatrix() {
    for (i in 0 until ledControl.getDeviceCount()) {
        ledControl.setIntensity(i, 1)
        ledControl.shutdown(i, false)
        ledControl.clearDisplay(i)
    }
}
```

This is a generic `init` method that iterates over all the chained devices (just one in our case) and, for each one of them, sets the intensity to the minimum (intensity varies from 0 to 15), sets shutdown to `false` -equivalent to enable it-, and clears the display.

Finally, let's take a look at `swapSprite`:

```
private var currentSprite = 0

private fun swapSprite() {
    currentSprite++
    for (row in 0..7) {
        ledControl.setRow(0, row, sprites[currentSprite % sprites.size][row])
    }
}
```

We increment the `currentSprite` variable and then use the module operator to get the correct one from the array of sprites.

Note that we use the `setRow` method that receives the device position and a byte that represents the values. The driver also offers an equivalent method to set a column and another one to set an individual pixel.

 This driver provides methods to set a row, a column, and a single pixel on the matrix.

The only missing part is the array of sprites, which, in our case, will be just two from the same model of `Space Invaders`:

```
val sprites = arrayOf(
        byteArrayOf(
            0b00011000.toByte(),
            0b00111100.toByte(),
            0b01111110.toByte(),
            0b11011011.toByte(),
            0b11111111.toByte(),
            0b00100100.toByte(),
            0b01011010.toByte(),
            0b10100101.toByte()
        ),
        byteArrayOf (
            0b00011000.toByte(),
            0b00111100.toByte(),
            0b01111110.toByte(),
            0b11011011.toByte(),
            0b11111111.toByte(),
            0b01011010.toByte(),
            0b10000001.toByte(),
            0b01000010.toByte()
        )
    )
```

That is just a binary definition of the sprites for positions 0 and 1, but, given that we are using rows, it allows us to visualize the sprite quite well. Feel free to play around, replace them, and add more elements to the array to animate other characters.

SPI version of SSD1306

In the previous chapter, we used the SSD1306 display via I2C. Now we are going to use it with SPI. In this case, we can have faster and more reliable communication at the expense of a more complex protocol.

It is quite common that a peripheral can work over SPI as well as over I2C. Most screens and sensors have this option. Note that SPI, although faster, has a limited number of slaves and also requires more wiring.

 Many components are designed to be used either over SPI or I2C. We are going to look at the screen to have one example.

In principle you should only have one or two extra pins, but in some cases the board has a few more. Let's look at the different pins you can find. Note that the labels on the pins may vary a lot between manufacturers; however, the pins are always the same:

- **Vcc**: to be connected to 5v.
- **Ground, GND**: to be connected to Ground.
- **SCL, CLK, SCK, or D0**: to be connected to the SPI clock.
- **MOSI, SDA, or D1**: to be connected to the SPI data (MOSI).
- **DC, D/C, or A0**: data/control pin. Used to tell the display whether we are sending data (pixels) or control (configuration). To be connected to any GPIO.
- **RS, RES, RST, or RESET**: reset pin. To be connected to any GPIO. It is used to reset the configuration of the display during initialization. This process may take a few seconds.
- **CS (may not be present)**: Chip select. To be connected to the chip select of the bus, either CE0 or CE1.

 The driver does reset the display on initialization; that is a synchronous action that takes a few seconds.

It is very important to note that some breakout circuits for this display do not have the CS pin exposed. If that is the case for your component, it will only work if it is the only peripheral on the SPI bus.

> Note that some breakout circuits may not expose the CS pin. If you have one of those, you have to keep it alone on the SPI bus.

The official driver from Google only covers the I2C wiring; to use SPI we need to use a different one:

```
dependencies {
    [...]
    implementation 'com.plattysoft.things:ssd1306:+'
}
```

And then we make an equivalent example as the one for I2C:

```
val display = Ssd1306.openSpi(SPI_BUS, DC_PIN, RS_PIN, 128, 64)
for (i in 0 until display.lcdWidth) {
    for (j in 0 until display.lcdHeight) {
        display.setPixel(i, j, i % display.lcdHeight > j)
    }
}
// Render the pixel data
display.show()
// Cleanup
display.close()
```

Note that we are using constants for the bus and the pin names instead of `BoardDefaults` just for simplicity.

The only difference with the I2C driver is the method to open the device. It now requires extra parameters for the DC and RS pins, as well as optional width and height (default display size is 128x64). The rest of the code is exactly the same as for I2C.

Summary

In this chapter we have learned about the SPI protocol and its similarities and differences with I2C. We have also checked how to use the RGD LED strip of the Rainbow HAT without the meta driver and we have seen how to work with a couple of screen components that are wired over SPI: LED matrix and OLED screens.

Now that we have completed the review of the basic protocols for communicating with peripherals, let's explore some areas where Android Things really sets itself on a different level in terms of libraries and utilities that allow us to build awesome IoT projects. It is time to look into the real power of Android Things.

The Real Power of Android Things

In this final chapter, we are going to explore some areas where Android Things really sets itself apart from other IoT platforms. This happens when we leverage on Android libraries and services that make building better and more compelling projects easier. The areas we are going to look at are:

- Using Android UI
- Companion apps and communication
- Other cool stuff

Technical requirements

You will be required to have Android Studio and Android Things installed on a developer kit. You also will require many hardware components to effectively perform the tasks given in this chapter. The components are very interesting to have, just to see them working, but the Rainbow HAT is particularly important. We go into details about the developer kits and how to pick the right one, as a part of Chapter 1, *Introducing Android Things*. Finally, to use the Git repository of this book, you need to install Git.

The code files of this chapter can be found on GitHub:
https://github.com/PacktPublishing/Android-Things-Quick-Start-Guide.

Check out the following video to see the code in action:

http://bit.ly/2PSHOaA.

Using Android UI

As we stated in the introductory chapter, displays are optional on Android Things by design, but that does not mean that we cannot have them. Given that the iMX7D developer kit includes a touchscreen and the Raspberry Pi has an HDMI output, giving the user access to this UI should be reasonably simple.

We can indeed create a UI, and what's even more interesting is that we can use the Android framework to build it. That allows us to reuse our knowledge of views and layouts to build beautiful UIs for our IoT projects; even animations if we want to.

 All the UI elements from the Android SDK are available on Android Things.

As a simple example, we are going to build an app that has a toggle control to change the status of the red LED and that displays the temperature from the sensor. With this example we show the flow from the UI to the peripherals (LED), and from the peripherals to the UI (temperature). We will be reusing this concept in the section about companion apps.

Feel free to create your own XML layout. The one in the code sample looks like this:

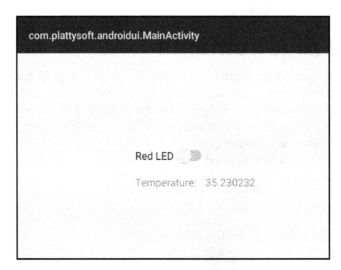

The screen is just a `RelativeLayout` with a `Switch` centered and two `TextViews` below it for the temperature: one for the label and one for the value.

This layout is basic Android knowledge, and teaching Android UI is beyond the scope of this book. If you are not familiar with it, there are plenty of examples and tutorials online, and even other books.

The activity code is as follows:

```
class MainActivity : Activity() {

    lateinit var redLed: Gpio
    lateinit var temperatureSensor : Bmx280

    override fun onCreate(savedInstanceState: Bundle?) {
        super.onCreate(savedInstanceState)
        setContentView(R.layout.activity_main)

        redLed = RainbowHat.openLedRed()

        temperatureSensor = RainbowHat.openSensor()
        temperatureSensor.setMode(Bmx280.MODE_NORMAL)
        temperatureSensor.temperatureOversampling = Bmx280.OVERSAMPLING_1X

        configureLedSwitch()

        configureTemperatureReading()
    }

    override fun onDestroy() {
        super.onDestroy()
        redLed.close()
        temperatureSensor.close()
    }

    [...]
}
```

We have the standard initialization and cleanup for the LED and the sensor; we also have two extra methods to configure the LED switch and the temperature reading, which we will be looking at separately:

```
private fun configureLedSwitch() {
    redLedSwitch.setOnCheckedChangeListener{
        compoundButton: CompoundButton, b: Boolean ->
        redLed.value = b
    }
}
```

The `redLedSwitch` variable is a `Switch`, so we can use a `CheckedChangeListener`. Once we add the listener, the callback receives a `CompoundButton` parameter representing the switch and a `Boolean` representing the new state. Since this listener is only attached to the red LED switch, we do not need to filter the events based on the source.

Inside the callback we just need to change the value of the `redLed` to the new value of the switch. Note that since the LED is a peripheral, we do not need to use the UI thread to modify it.

Let's look at the temperature reading:

```
private val handler = Handler(Looper.getMainLooper())

private fun configureTemperatureReading() {
    handler.post(object : Runnable {
        override fun run() {
            val temperature = temperatureSensor.readTemperature()
            temperatureValue.text = temperature.toString()
            handler.postDelayed(this, 1000)
        }
    })
}
```

This case is a bit trickier. We use the typical handler/runnable approach to have the temperature evaluated every second, and then we just set the value on the `temperatureValue` text view.

Note that we passed the main looper as a parameter when we constructed the looper, so all the calls are on the UI thread already. If we were to be reading the temperature on another thread, we would need to send a runnable to the main thread to be able to modify the content of the `TextView`.

As you see, adding a UI to our IoT device is quite simple.

Companion apps and communication

One of the most typical setups for IoT involves two apps: one for the *thing* and another one for *mobile*. It just makes sense to control your IoT device from your phone. With Android Things and Android Studio we can create both apps under the same project, each one being its own module. This setup also allows us to have another *common* library module that is used by both including areas such as a communication data model.

With Android Studio, we can have a mobile app and a Things app as modules of the same project

The big question about having a companion app is about how it should communicate with the IoT device. There are multiple ways to do so. The following three are the most common:

- Hosting a REST API server on the things app and a client on the mobile one
- Firebase Realtime Database integration on both sides
- Nearby communication

Each option has advantages and disadvantages. In this section we will be exploring the broad setup for each of them.

We will extend the previous example of temperature and red LED and we will split it into two parts: the control of the peripherals inside the things module and the UI on the mobile one.

In all the examples, the things apps will have the same block of initialization and cleanup for the peripherals, which is the same as for the previous example, and the mobile apps will have the same layout, which will also be the same as for the previous example.

With that settled, let's make a REST API.

REST API using NanoHttpd/Retrofit

The simplest and most standard way to interact with an IoT device is to provide a REST API. To do that, we need to include a web server in our things app and implement the handling of the requests. A simple library that helps us to achieve that easily is `NanoHttpd`. To include it we need to add the following dependency to our `things` module:

```
dependencies {
    [...]
    implementation 'org.nanohttpd:nanohttpd:2.2.0'
}
```

The architecture we will use is quite simple. The activity will create a web server that will notify the requests received via an interface that represents the methods exposed on the API. In our case the interface is:

```
interface ApiListener {
    fun onGetTemperature(): JSONObject
```

```
    fun onPostLed(request: JSONObject)
}
```

We have one GET method for the temperature value that returns the information encoded in a JSONObject and a POST method to change the LED status, which also receives the request body as a JSONObject.

Let's look at the activity:

```
class MainActivity : Activity(), ApiListener {
    [...]
    lateinit var apiServer: ApiServer

    override fun onCreate(savedInstanceState: Bundle?) {
        super.onCreate(savedInstanceState)
        [...]
        apiServer = ApiServer(this)
    }

    override fun onDestroy() {
        super.onDestroy()
        apiServer.stop()
        [...]
    }
    [...]
}
```

We can see that the activity implements ApiListener and creates a server inside onStart and it stops it inside onStop.

Next, let's see how we implement the ApiListener methods in the activity:

```
override fun onGetTemperature(): JSONObject {
    val response = JSONObject()
    val value = temperatureSensor.readTemperature().toDouble()
    response.put("temperature", value)
    return response
}

override fun onPostLed(request: JSONObject) {
    val status = request.get("status") as Boolean
    redLed.value = status
}
```

To return the temperature, we create a JSONObject and we add a field that is the temperature as a double. Then, we return that object to be used as the response body.

To modify the value of the LED, we get the field named `status` from the request object and we set it to the value of the `redLed`.

The `ApiServer` class is just a simple implementation of `NanoHttp` where we start it on port 8080 and listen for the proper methods and path, in our case, `GET /temperature` and `POST /led`:

```
class ApiServer(private val listener: ApiListener) : NanoHTTPD(8080) {

    init {
        start(NanoHTTPD.SOCKET_READ_TIMEOUT, false)
    }

    override fun serve(session: IHTTPSession): Response {
        val path = session.uri
        val method = session.method

        if (method === Method.GET && path == "/temperature") {
            val response = listener.onGetTemperature()
            return newFixedLengthResponse(response.toString())
        }
        else if (method === Method.POST && path == "/led") {
            val request = getBodyAsJson(session)
            listener.onPostLed(request)
            return newFixedLengthResponse("")
        }
        return newFixedLengthResponse(Response.Status.NOT_FOUND,
            NanoHTTPD.MIME_PLAINTEXT,
            "Error 404, file not found.")
    }
}
```

Where `getBodyAsJson` is just a utility method to keep the code more concise:

```
private fun getBodyAsJson(session: NanoHTTPD.IHTTPSession): JSONObject {
    val files = HashMap<String, String>()
    session.parseBody(files)
    var content = files["postData"]
    return JSONObject(content)
}
```

We can test this straight away -without building the companion app- using a browser with something like Restlet or Postman.

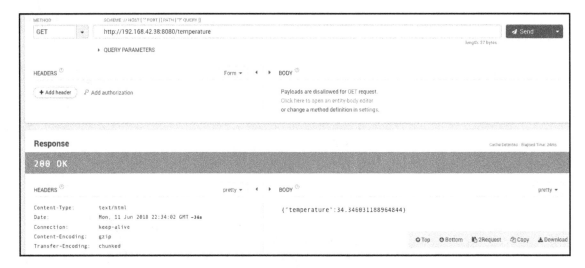

The api-mobile module can be built using the retrofit library and it is included in the examples so you can check it. But since it is a very standard Android example we won't go into that part here.

Retrofit is a very useful library to make REST clients. For more info you can check the official page: `http://square.github.io/retrofit/`.

Using a REST API has some clear benefits: it is a simple setup and a standard way of communicating (that is, if you want to open your IoT thing to other developers). It also comes with a few challenges, mostly related to IP: you need to know the IP of the device to connect to and you need to either expose it to the internet or only work on a LAN. Also, if you want any type of authentication, you need to build it yourself.

Firebase Realtime Database

The second approach we are going to explore is Firebase Realtime Database. This is a very common example because it showcases how third-party tools can simplify and speed up your development.

For this part, we will have a shared realtime database used by both the things and the mobile apps. That database will have one row with the current values of the temperature and the LED status.

The things app will write the temperature entry to the database and listen for modifications on the LED one.

The mobile app will do the reverse operation: write the LED value when the switch changes and listen for changes on the temperature value to display them on the UI.

To begin with, we need to create a project on the Firebase console and add a realtime database. We can do that at `https://console.firebase.google.com/`.

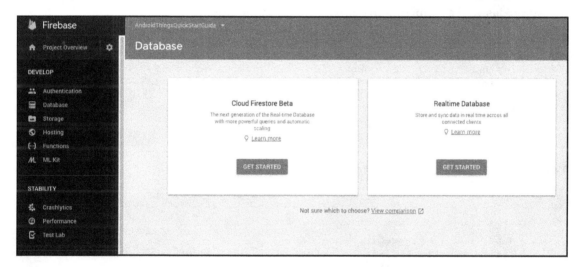

To keep it simple and get things moving, we will start in test mode, but you should take care of authentication if you plan to use this for anything serious. Test mode is just for trying out how it works.

Do not forget to remove test mode and use authentication if you plan to use Firebase on a real-life scenario.

The next step is to add the Android apps (things and mobile) to your Firebase project. Once you do that, you can download a `google-services.json` file that is generated with the information of the app and the Firebase project. We need to add that file to both of our modules.

With this, the backend part is completed.

Now, to enable Firebase in our project, we need to add Google Services to the classpath. We do that on the project level build.gradle:

```
buildscript {
  dependencies {
    [...]
    classpath 'com.google.gms:google-services:4.0.1'
  }
}
```

And then, on each module, we add the dependencies to firebase-core and firebase-database, as well as applying the google-services plugin:

```
dependencies {
    [...]
    implementation 'com.google.firebase:firebase-core:16.0.0'
    implementation 'com.google.firebase:firebase-database:16.0.1'
}
[...]
// Add to the bottom of the file
apply plugin: 'com.google.gms.google-services'
```

With the setup completed, let's start with the things module:

```
lateinit var firebaseReference: DatabaseReference

override fun onCreate(savedInstanceState: Bundle?) {
    super.onCreate(savedInstanceState)
    FirebaseApp.initializeApp(this)
    val database = FirebaseDatabase.getInstance()
    firebaseReference = database.getReference()

    [...]

    configureTemperatureReading()
    configureLedSwitch()
}
```

We initialize the firebase app and then get a reference to the instance of the database. This reference is what we will use to write values and configure listeners.

We configure the temperature reading in the typical fashion of handler/runnable:

```
private fun configureTemperatureReading() {
    handler.post(object : Runnable {
        override fun run() {
            val temperature = temperatureSensor.readTemperature()
            firebaseReference.child("temperature").setValue(temperature)
            handler.postDelayed(this, 1000)
        }
    })
}
```

Every time we read a new temperature value, we set the value to the child named temperature on the root of the database reference. This is very similar to creating a JSON object in the previous example, except that this time we don't have to pass it anywhere; Firebase will handle it for us in the background.

Reading the LED is a bit more complex:

```
private fun configureLedSwitch() {
    firebaseReference.child("redLED").addValueEventListener (
        object: ValueEventListener {
            override fun onCancelled(p0: DatabaseError) {
                // Nothing to do here
            }
            override fun onDataChange(snapshot: DataSnapshot) {
                val redLedState = snapshot.getValue(Boolean::class.java)
                if (redLedState != null) {
                    redLed.value = redLedState
                }
            }
        })
}
```

We add a ValueEventListener to the child named redLED. When the data of that value changes, we receive a snapshot of that data. From this snapshot, we can read the value of the Boolean and update the value of the redLed variable.

If we now go to the **Database** section on the Firebase console, we can check what it looks like. Since we have not written the value for the redLED, that field does not exist, but if we add it manually on the console and change it, the things app will react to it.

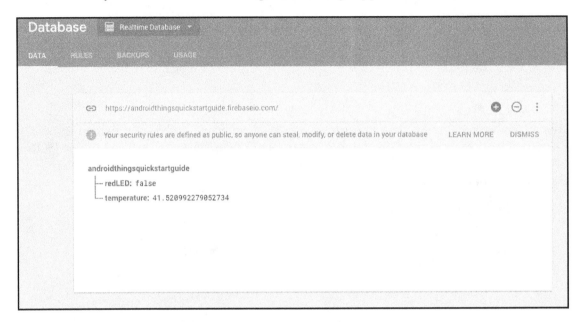

Don't forget to get internet connectivity to your dev kit (especially if you are using the iMX7D).

Now, on the mobile side we have a similar logic, although reversed: the `redLed` switch writes and the temperature has a listener:

```
lateinit var firebaseReference: DatabaseReference

override fun onCreate(savedInstanceState: Bundle?) {
    super.onCreate(savedInstanceState)
    setContentView(R.layout.activity_main)

    FirebaseApp.initializeApp(this)
    val database = FirebaseDatabase.getInstance()
    firebaseReference = database.getReference()

    redLedSwitch.setOnCheckedChangeListener {
        compoundButton: CompoundButton, b: Boolean ->
        firebaseReference.child("redLED").setValue(b)
```

```
        }
    }
```

The initialization of Firebase is also the same as for the things app.

Adding a `ValueEventListener` and modifying the `TextView` temperature should be straightforward given everything we have done in this chapter so far, so it is left as an exercise for the reader. However, it is solved in the code examples, in case you want to look it up.

Firebase is extremely simple to set up, and it is not restricted to your local network. You can control your IoT device from anywhere in the world since Firebase is cloud-based. The only drawback of it is that it ties you to a particular service (Firebase) that is outside your control.

Nearby

Nearby is a library made by Google that allows us to create features based on proximity. It uses mostly Bluetooth, but also Wi-Fi. It has three main modes: connections, messages, and notifications. Of the three modes, only connections is fully supported on Android Things 1.0.1.

Nearby Connections is a peer-to-peer networking API that allows apps to easily advertise, discover, connect to, and exchange data with nearby devices in real time, regardless of network connectivity.

To use Nearby, we need to add the following dependency in both modules: mobile and things:

```
dependencies {
    [...]
    implementation 'com.google.android.gms:play-services-nearby:15.0.1'
}
```

The following permissions are required in your `AndroidManifest` for both modules:

```
<uses-permission android:name="android.permission.BLUETOOTH" />
<uses-permission android:name="android.permission.BLUETOOTH_ADMIN" />
<uses-permission android:name="android.permission.ACCESS_WIFI_STATE" />
<uses-permission android:name="android.permission.CHANGE_WIFI_STATE" />
<uses-permission android:name="android.permission.ACCESS_COARSE_LOCATION"
/>
```

There are also a few constants that we are going to use, so having a companion object makes sense:

```
companion object {
    private val TAG = MainActivity::class.java.simpleName

    private const val SERVICE_ID = "com.plattysoft.nearby"
    private const val NICKNAME = "things"
    private val STRATEGY = Strategy.P2P_STAR
}
```

Service id is a unique identifier of your service; it is usually the package name. It is used by the clients to filter out all the service ids that they are not interested in.

 Nearby connects apps that use the same service id.

Nickname is a human-friendly description of the device. This should be different on the mobile and things modules.

Finally, **strategy** defines if we want to use a star or a cluster. For more information about the characteristics of each topology, you can visit the official page at `https://developers.google.com/nearby/connections/strategies`.

And then, since we are going to use Connections, we need to get the connections client, which can be obtained via `Nearby.getConnectionsClient`.

```
nearbyConnections = Nearby.getConnectionsClient(this)
```

Now that we have the basic requirements set up for both modules, we can explore the procedure. We will make the things app advertise and the mobile app discover.

There is a first step where one app is advertising and another one is discovering. When that happens, the things endpoint is discovered by the mobile one and it will initiate a connection. The things app needs to approve the connection, and so does the mobile app.

Once the connection is established, each endpoint can send payloads to the other, using an internal endpoint id.

So, let's look at each part separately.

Advertising and discovering

To create a nearby connection, a device needs to advertise itself so other nearby devices that are discovering can find it and establish a connection. Remember that both endpoints need to use the same service id and strategy.

The code to start advertising is as follows:

```
nearbyConnections.startAdvertising(
        NICKNAME,
        SERVICE_ID,
        connectionLifecycleCallback,
        AdvertisingOptions.Builder().setStrategy(STRATEGY).build())
```

Note that you can set success and failure listeners. And you should do that if you use this technique in production. We have left them out of the example to keep the code simple.

On the other hand, a device can start discovery mode in a very similar way:

```
nearbyConnection.startDiscovery(
        SERVICE_ID,
        endpointDiscoveryCallback,
        DiscoveryOptions.Builder().setStrategy(STRATEGY).build())
```

Note that we have a parameter that is an `EndpointDiscoveryCallback` instead of the `ConnectionLifecycleCallback` we had for advertising. This callback is what will be used to notify us when a new endpoint is discovered.

Initiating and confirming connections

A connection can be initiated as soon as we discover a new endpoint. That is done inside the `endpointDiscoveryCallback` on the mobile module:

```
private val endpointDiscoveryCallback = object :
EndpointDiscoveryCallback() {
    override fun onEndpointFound(endpointId: String,
discoveredEndpointInfo: DiscoveredEndpointInfo) {
        connectToEndpoint(endpointId)
    }
    override fun onEndpointLost(endpointId: String) {
        // A previously discovered endpoint has gone away.
    }
}
```

On the callback we receive the endpoint id and the endpoint info. To initiate a connection we only need the endpoint id.

```
private fun connectToEndpoint(endpointId: String) {
    nearbyConnection.requestConnection(
            NICKNAME,
            endpointId,
            connectionLifecycleCallback)
}
```

Note that we are passing a `connectionLifecycleCallback` to `requestConnection`. That callback is the one that will be invoked on the discovery endpoint, and we already had a callback of this type as a parameter to `startAdvertising`, so both sides use the same interface.

Once the connection is requested, both endpoints are required to accept it. An authentication token is provided that can be manually confirmed. For simplicity, we are just going to auto confirm on both ends.

```
private val connectionLifecycleCallback = object :
ConnectionLifecycleCallback(){
    override fun onConnectionResult(endpointId: String, result:
ConnectionResolution) {
        if (result.status.statusCode == STATUS_OK) {
            connectedEndpoint = endpointId
        }
    }
    override fun onDisconnected(endpointId: String) {
        if (endpointId.equals(connectedEndpoint)) {
            connectedEndpoint = null
        }
    }
    override fun onConnectionInitiated(endpointId: String, connectionInfo:
ConnectionInfo) {
        nearbyConnection.acceptConnection(endpointId, payloadCallback)
    }
}
```

We are keeping one single endpoint id as a class variable, we assign it when the connection is completed and we unassign it when the connection is lost. This is fine for our single discoverer case, but for a proper implementation we should keep a list of connected endpoints.

Finally, every time a connection is initiated, we accept it and assign a `payloadCallback` to be notified of incoming payloads.

Sending and receiving data

On the things side, we are going to send the temperature every second. To do so, we will encode it inside a JSONObject, using the same technique as for the REST API. In this case we create a Payload from the bytes of the JSONObject once it is converted to a string and we send that to the connectedEndpoint. Please note that we will only invoke this method if the connectedEndpoint is already set.

```
private fun sendTemperatureToNearby(temp: Float) {
    val json = JSONObject()
    json.put("temperature", temp)
    val payload = Payload.fromBytes(json.toString().toByteArray())
    nearbyConnections.sendPayload(connectedEndpoint!!, payload)
}
```

And, in a straightforward way, the payload will arrive on the other device via the payloadCallback we passed in when accepting the connection:

```
private val payloadCallback = object: PayloadCallback() {
    override fun onPayloadReceived(p0: String, payload: Payload) {
        // Info about the temperature
        val json = payload.getAsJson()
        runOnUiThread {
            val value = json.getDouble("temperature").toString()
            findViewById<TextView>(R.id.temperatureValue).setText(value)
        }
    }

    override fun onPayloadTransferUpdate(p0: String, p1:
PayloadTransferUpdate) {
        // Nothing to do here
    }
}
```

We just convert the payload to JSON, get the right field, and post a runnable to the UI thread to update the UI. Quite simple.

Note that we have created an extension function that gets the payload as a JSONObject. It is just a couple of lines, but makes the code much more readable.

```
private fun Payload.getAsJson(): JSONObject {
    val string = String(this.asBytes()!!)
    return JSONObject(string)
}
```

Similarly to what we just saw, we can send a payload from mobile to things to change the state of the LED. As in the case of the REST API, this is solved in the code samples, but it is left as an exercise for the reader.

Given that Nearby does not require connectivity -it is a mesh network-, it is very handy to communicate with devices that are nearby. The advertising and discovery steps solve the problem of knowing the IP we had on the REST API approach.

However, there are a couple of drawbacks. First, it does have a lot of states and managing all of them properly can be complex, including the case when Bluetooth is not enabled, or if any of the required permissions are not granted.

There is also the lack of structure of the payloads. You can do whatever you want, which is good because it gives you flexibility, but it also requires you to define clearly the states and communications your apps perform.

More cool stuff

A bit more briefly, let's look at other interesting options that Android Things offers us.

TensorFlow – image classifier

TensorFlow is a library provided by Google to do machine learning. You can do the processing on the cloud, or you can do it on the device itself using TensorFlow Lite, which is when it gets interesting.

One of the most common use cases is image classifiers, where we train a model to recognize different categories of images and then use it to classify new ones. There are two very interesting codelabs, one on how to integrate an image classifier on Android Things and another one about how to train you own classifier.

You can learn how to integrate TensorFlow Lite on Android Things with the codelab at `https://codelabs.developers.google.com/codelabs/androidthings-classifier/`.

The first codelab goes over the following topics:

- How to integrate the TensorFlow Lite inference library for Android
- How to feed data into a TensorFlow model
- How to capture an image using the Android camera APIs

Google provides a default configuration of TensorFlow Lite as a pre-compiled AAR on JCenter, so you can start using it in your project by simply adding a dependency line to your `build.gradle`, instead of installing and configuring a lot of native compilation tools:

```
dependencies {
    [...]
    implementation 'org.tensorflow:tensorflow-lite:0.1.7'
}
```

TensorFlow uses the camera, and Android Things provides the same Camera APIs that a phone does. However, using cameras on Android involves some non-trivial code, due to the optimized pipeline used by the camera hardware and the image surface. This codelab has built a few utility classes to simplify those tasks.

You can learn how to train your own image classifier on this codelab at `https://codelabs.developers.google.com/codelabs/tensorflow-for-poets/`.

Once you have done the integration, you can move on and create your own model, which, ultimately just results in a file that you add to the project that contains the neural network.

Publish subscribe buses

Another typical communication method for IoT is the use of a publish subscribe bus, where the IoT devices push information as a stream. These buses are hosted on the cloud and you can use them to build your own services on top of them.

Conceptually, it is not too different from how Firebase Realtime Database works. It is a bit more structured, since it follows a well- known architecture pattern.

There are several companies that offer this service, but the most well-known ones are Google's Cloud Pub/Sub and PubNub.

Both are based on the same concepts: the IoT device registers with the service, opens a channel with publishing permission, and then it publishes the sensor updates to it while other services will subscribe to this channel, read the updates, and maybe decide to act on it.

You can also have the IoT device subscribe to its own channel (or stream) and get notifications when it is required to act on some event, or you can configure a separate channel for that.

> Google also offers `Cloud IoT Core` as a more complete solution than just Pub/Sub.

Publish subscribe buses are complex, full of features, and also beyond the scope of this book. To learn more about them, you can check the official pages:

- Google Cloud Pub/Sub: `https://cloud.google.com/pubsub/`
- PubNub: `https://www.pubnub.com/`

Google Assistant – Home control

Google Assistant is another very interesting integration. With this, you can make your IoT device react to a voice command to your phone or to your Google Home.

There are several ways to achieve this interaction, from creating your own agent to doing a smart home integration directly.

> You can check the smart home integration with Google Assistant here: `https://developers.google.com/actions/smarthome/`

In any case, most of the code required to do this falls into the backend side and the setup is slightly complex. If you are interested in exploring this, I recommend you to follow the instructions from the official example to set the backend from here: `https://github.com/actions-on-google/smart-home-nodejs`

Online resources

Finally, there are some relevant online resources for developers interested in Android Things and IoT.

The most obvious one is **Build with Android Things** (`https://androidthings.withgoogle.com/`), which is directly managed by Google and includes featured community projects, lists of drivers, samples, and snippets to get you started on many different aspects of Android Things.

The other most interesting resource is **hackster.io**, which is a maker's community where people share projects. They have an area for Android Things, which you can visit at `https://www.hackster.io/google/products/android-things`. Some particular favorites of mine are:

- `DrawBot`, a DIY robotic drawing robot that can take your selfie and sketch your portrait within minutes.
- `Candy Dispenser`, a game that asks the user for a specific thing like a bird, dog, or cat, and the user should show a photo of that thing in the predefined time to win candies.
- `Flos Affectus`, a cluster of robotic flowers that *bloom* and *un-bloom*, depending on the expression detected on the user's face.

Summary

And that's a wrap. We started the book by learning the basic concepts of Android Things and how to set up a developer kit to then use the components of the Rainbow HAT, explore the different protocols that are used, and play with other components that are commonplace in IoT projects.

In this last chapter, we took a glimpse into what makes Android Things really powerful (besides the security updates policy) and that is all the libraries and resources that are at our fingertips. We have explored just a few, from adding UI to our devices to simple methods for communicating with a companion app or even a backend service. We also peeked at image classifiers and voice assistants. Finally, we presented some interesting online resources.

By now you should be able to find your way into any IoT project that you had in mind and make it a reality with Android Things.

Happy hacking & tinkering!

Pinouts diagrams and libraries

Raspberry Pi Pinout

The following pinout diagram illustrates the peripheral I/O interfaces of the Raspberry Pi 3 developer kit:

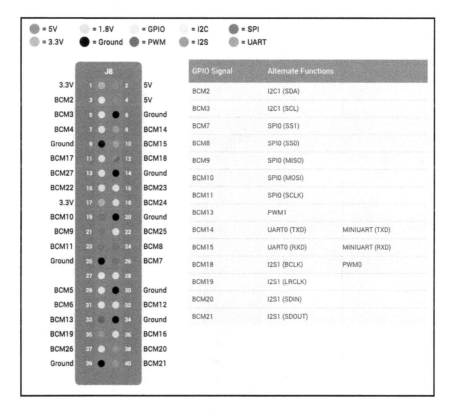

Pinout diagram from the official Android Things developer

page: https://developer.android.com/things/hardware/raspberrypi-io.

NXP iMX7D Pinout

The following pinout diagram illustrates the peripheral I/O interfaces of the iMX7D developer kit:

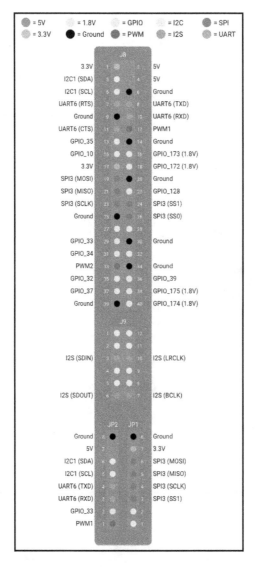

Pinout diagram from the official Android Things developer

page: https://developer.android.com/things/hardware/imx7d-pico-io

Supported and unsupported libraries

This section describes the state of unsupported Android features and intents on Android Things 1.0, as well as the state of available and unavailable Google APIs on Android Things.

Unsupported features

Given that Android Things is optimized for embedded devices that may not contain the same feature set as an Android phone or tablet, not all the APIs are present. In most cases, this comes from the restriction of not requiring a UI. This table lists the features from Android that are not supported by Android Things and the affected framework APIs:

Feature	API
System UI *(status bar, navigation buttons, quick settings)*	NotificationManager KeyguardManager WallpaperManager
VoiceInteractionService	SpeechRecognizer
android.hardware.fingerprint	FingerprintManager
android.hardware.nfc	NfcManager
android.hardware.telephony	SmsManager TelephonyManager
android.hardware.usb.accessory	UsbAccessory
android.hardware.wifi.aware	WifiAwareManager
android.software.app_widgets	AppWidgetManager
android.software.autofill	AutofillManager
android.software.backup	BackupManager
android.software.companion_device_setup	CompanionDeviceManager
android.software.picture_in_picture	Activity Picture-in-picture
android.software.print	PrintManager
android.software.sip	SipManager

You can use `hasSystemFeature()` to determine whether a given device feature is supported in runtime.

Common intents and content providers

Android Things lacks some system apps and system providers that are usually installed on Android. This is mostly related to the fact that they require a UI to work, or that they do not make much sense in the realm of IoT. Avoid using common intents as well as the following content provider APIs in your apps:

- CalendarContract
- ContactsContract
- DocumentsContract
- DownloadManager
- MediaStore
- Settings
- Telephony
- UserDictionary
- VoicemailContract

There are also the Google APIs for Android, most of which are supported on Android Things.

Available Google APIs

Although not all Google APIs are supported on Android Things, there is a good amount that are, and they provide very interesting functionalities. The Google APIs that are currently supported on Android Things are as follows:

- Awareness
- Cast
- Google Analytics for Firebase
- Firebase Authentication (does not include the open source FirebaseUI Auth component)
- **Firebase Cloud Messaging** (**FCM**)
- Firebase Crash Reporting
- Firebase Realtime Database
- Firebase Remote Config

- Firebase Storage
- Fit
- Instance ID
- Location
- Maps
- Nearby Connections
- Nearby Messages (does not include modes that require the user content dialog)
- Places
- Mobile Vision
- SafetyNet

Unavailable Google APIs

Again, most of the unsupported APIs rely heavily on the UI, so they are absent. Note that Firebase Notifications is not available, but Firebase Cloud Messaging is, which you can use to send and receive messages to trigger actions on your phone:

- AdMob
- Google Pay
- Drive
- Firebase App Indexing
- Firebase Dynamic Links
- Firebase Invites
- Firebase notifications
- Play fames
- Sign-in

Other Books You May Enjoy

If you enjoyed this book, you may be interested in these other books by Packt:

Android Things Projects
Francesco Azzola

ISBN: 9781787289246

- Understand IoT ecosystem and the Android Things role
- See the Android Things framework: installation, environment, SDK, and APIs
- See how to effectively use sensors (GPIO and I2C Bus)
- Integrate Android Things with IoT cloud platforms
- Create practical IoT projects using Android Things
- Integrate Android Things with other systems using standard IoT protocols
- Use Android Things in IoT projects

Robot Operating System Cookbook
Kumar Bipin

ISBN: 9781783987443

- Explore advanced concepts, such as ROS pluginlib, nodelets, and actionlib
- Work with ROS visualization, profiling, and debugging tools
- Gain experience in robot modeling and simulation using Gazebo
- Understand the ROS Navigation Stack for mobile robots
- Configure a MoveIt! package for a manipulator robot
- Develop an autonomous navigation framework for MAV using ORB SLAM and MoveIt
- Integrate sensors, actuators, and robots into the ROS ecosystem
- Get acquainted with the ROS-Industrial package with hardware support, capabilities, and applications

Leave a review - let other readers know what you think

Please share your thoughts on this book with others by leaving a review on the site that you bought it from. If you purchased the book from Amazon, please leave us an honest review on this book's Amazon page. This is vital so that other potential readers can see and use your unbiased opinion to make purchasing decisions, we can understand what our customers think about our products, and our authors can see your feedback on the title that they have worked with Packt to create. It will only take a few minutes of your time, but is valuable to other potential customers, our authors, and Packt. Thank you!

Index